Pocket Guide to Mycological Diagnosis

T0094143

Pocket Guides to Biomedical Sciences

https://www.crcpress.com/Pocket-Guides-to-Biomedical-Sciences/book-series/
CRCPOCGUITOB

The **Pocket Guides to Biomedical Sciences** series is designed to provide concise, state-of-the-art, and authoritative coverage on topics that are of interest to undergraduate and graduate students of biomedical majors, health professionals with limited time to conduct their own searches, and the general public who are seeking reliable, trustworthy information in biomedical fields.

Pocket Guide to Mycological Diagnosis

Edited by
Rossana de Aguiar Cordeiro

CRC Press
Taylor & Francis Group
Boca Raton London New York

CRC Press is an imprint of the
Taylor & Francis Group, an **informa** business

CRC Press
Taylor & Francis Group
6000 Broken Sound Parkway NW, Suite 300
Boca Raton, FL 33487-2742

© 2020 by Taylor & Francis Group, LLC
CRC Press is an imprint of Taylor & Francis Group, an Informa business

No claim to original U.S. Government works

Printed on acid-free paper

International Standard Book Number-13: 978-1-138-05594-0 (Hardback)
978-1-138-05593-3 (Paperback)

Visit the Taylor & Francis Web site at
http://www.taylorandfrancis.com

and the CRC Press Web site at
http://www.crcpress.com

Contents

Preface

In the past three decades, the epidemiology of fungal infections has changed dramatically. Along with the AIDS pandemic, extensive medical progress has led to an increase in the global population of severely immunocompromised patients, who are vulnerable to fungal pathogens. Since proper diagnosis is considered a pivotal factor for patient management, routine microbiology laboratories have been overloaded with requests for mycological tests. This situation has compelled many microbiologists to seek ongoing education in medical mycology—a field of knowledge neglected in many health courses.

In addition, the science of medical mycology has witnessed tremendous changes since the 1980s. The description of new species and species complexes, the discovery of fungal biofilms and their importance in pathogenesis, the continuous reports of refractory infections and antifungal resistance, and the increasing list of opportunistic fungal species have highlighted the importance of mycological diagnosis.

Cellular and molecular techniques, immunological methods, and more accurate microscopy equipment are now available to mycology laboratories. Furthermore, information regarding medical mycology, including identification of specific fungal pathogens, is widely available on the World Wide Web. Mycologists have to face the challenge of systematizing all this body of knowledge and applying it in routine diagnosis.

Therefore, this book aims to provide concise and useful information for microbiologists and professionals interested in the diagnosis of the most relevant fungal species of medical importance. Potential contributors were chosen based on their personal experience and previous authorship of widely cited publications on the subject.

We hope this book inspires young mycologists worldwide and contributes to enhancing the detection of fungal pathogens in routine laboratories.

Professor Rossana de Aguiar Cordeiro
Federal University of Ceará, Brazil

Editor

Rossana de Aguiar Cordeiro is a biologist and undertook her postgraduate training in Microbiology at the Federal University of Minas Gerais and Federal University of Ceará, both in Brazil. During the past two decades, she has dedicated her efforts to Medical Mycology, focusing on emerging fungal infections in Brazil, mainly coccidioidomycosis. She is the first author of more than 40 original research and review articles and, as an author, has contributed to the books *Molecular Detection of Human Fungal Pathogens* and *Manual of Security Sensitive Microbes and Toxins*, both published by CRC Press.

Contributors

Fernando Almeida-Silva
National Institute of Infectology
Fundação Oswaldo Cruz
Brazil

Silviane Praciano Bandeira
Specialized Medical Mycology Center
Department of Pathology and Legal Medicine
Federal University of Ceará
Brazil

Raimunda Sâmia Nogueira Brilhante
Specialized Medical Mycology Center
Department of Pathology and Legal Medicine
Federal University of Ceará
Brazil

Danielle Patrícia Cerqueira Macêdo
Department of Pharmaceutical Sciences
Center for Health Sciences, UFPE
Brazil

Carolina Maria da Silva
Department of Mycology
Center for Biosciences
Federal University of Pernambuco (UFPE)
Cidade Universitária
Recife, Brazil

Marcos de Abreu Almeida
National Institute of Infectology
Fundação Oswaldo Cruz
Brazil

Rossana de Aguiar Cordeiro
Specialized Medical Mycology Center
Department of Pathology and Legal Medicine
Federal University of Ceará
Brazil

João Nobrega de Almeida Júnior
Central Laboratory Division
Hospital das Clínicas
Faculty of Medicine
University of São Paulo
Brazil

Rodrigo de Almeida Paes
National Institute of Infectology
Fundação Oswaldo Cruz
Brazil

**André Luiz Cabral Monteiro de Azevedo
Santiago**
Department of Tropical Medicine
Center for Health Sciences, UFPE
Brazil

Zoilo Pires de Camargo
Laboratory of Emerging Fungal Pathogens,
   Cell Biology Division
Department of Microbiology, Immunology
   and Parasitology
Federal University of São Paulo
Brazil

Ana Maria Rabelo de Carvalho
Department of Mycology
Center for Biosciences
Federal University of Pernambuco (UFPE)
Cidade Universitária
Recife, Brazil

Mauro de Medeiros Muniz
National Institute of Infectology
Fundação Oswaldo Cruz
Brazil

Glaucia Morgana de Melo Guedes
Specialized Medical Mycology Center
Department of Pathology and Legal
   Medicine
Federal University of Ceará
Brazil

**Débora de Souza Colares Maia
Castelo-Branco**
Department of Pathology and Legal
   Medicine
Federal University of Ceará
Brazil

Márcia dos Santos Lazéra
National Institute of Infectology
Fundação Oswaldo Cruz
Brazil

Reginaldo Gonçalves de Lima-Neto
Department of Tropical Medicine
Center for Health Sciences, UFPE
and
Department of Mycology
Biosciences Center
Federal University of Pernambuco (UFPE)
Brazil

Rejane Pereira Neves
Department of Mycology
Center for Biosciences
Federal University of Pernambuco (UFPE)
Cidade Universitária
Recife, Brazil

Rosane Orofino-Costa
Dermatology Department
School of Medical Sciences
Rio de Janeiro State University
Brazil

Germana Costa Paixão
Center of Health Sciences
State University of Ceará
Brazil

Patrice Le Pape
IICiMed – EA 1155 Cibles et médicaments
    des infections de l'immunité et du cancer
University of Nantes
France

Claudia Vera Pizzini
National Institute of Infectology
Fundação Oswaldo Cruz
Brazil

Marcos Fábio Gadelha Rocha
Specialized Medical Mycology Center
Department of Pathology and Legal Medicine
Federal University of Ceará
and
School of Veterinary
State University of Ceará
Brazil

Anderson Messias Rodrigues
Laboratory of Emerging Fungal Pathogens,
    Cell Biology Division
Department of Microbiology, Immunology
    and Parasitology
Federal University of São Paulo
Brazil

José Júlio Costa Sidrim
Specialized Medical Mycology Center
Department of Pathology and Legal
    Medicine
Federal University of Ceará
Brazil

Luciana Trilles
National Institute of Infectology
Fundação Oswaldo Cruz
Brazil

Bodo Wanke
National Institute of Infectology
Fundação Oswaldo Cruz
Brazil

Rosely Maria Zancopé-Oliveira
National Institute of Infectology
Fundação Oswaldo Cruz
Brazil

List of Abbreviations

AD	atopic dermatitis
AdoMet/SAM	S-adenosylmethionine
AFLP	amplified fragment length polymorphism
ARF	ADP-ribosylation factor
ART	antiretroviral therapy
ASR	analyte-specific reagent
AST	Antimicrobial susceptibility testing
BAL	Bronchoalveolar lavage
BDG	1–3-β-D-glucan
BHI	brain heart infusion agar
BSL	biosafety level
CDC	Centers for Disease Control and Prevention
CF	complement fixation
CFW	calcofluor white fluorescent stain
CGB	l-canavanine glycine bromothymol blue
CLSI	Clinical Laboratory Standards Institute
CNS	central nervous system
CrAg	cryptococcal *capsular* polysaccharide *antigen*
CSF	*cerebrospinal fluid*
CVC	central venous catheters
DGGE	denaturing gradient gel electrophoresis
ECV or ECOFF	epidemiological cut-off value
EDTA	ethylenediamine tetraacetic *acid*
EIA	enzyme immunoassay
ELISA	enzyme-linked immunosorbent assay
EUCAST	European Committee on Antimicrobial Susceptibility Testing
FFPE	formalin-fixed paraffin-embedded tissue
FISH	fluorescence in situ hybridization
FITC	fluorescein isothiocyanate
FL	folliculitis
FTD	fast track diagnostics
GAPDH	glyceraldehyde 3-phosphate dehydrogenase
GMS	Gomori methenamine-silver nitrate *stain*
gp43	glycoprotein of 43,000 daltons (*Paracoccidioides* antigen)
GXM	capsular antigen glucuronoxylomannan from *Cryptococcus*
HE	hematoxylin and eosin
HIV	human immunodeficiency virus
HPA	*H. capsulatum* polysaccharide antigen
ID	immunodiffusion
IF	immunofluorescence
IGS	intergenic spacer-1
IRIS	immune reconstitution inflammatory syndrome
ISHAM	International Society of Human and Animal Mycology
ITS	internal transcribed spacer
KL-6	Krebs von den Lungen-6 antigen
KOH	potassium hydroxide
LA	latex agglutination test
LAMP	loop-mediated isothermal amplification
LDH	lactate dehydrogenase
LFA	lateral flow immunoassay
LSU	large subunit
MALDI-TOF MS	matrix-assisted laser desorption/ionization time-of-flight mass spectrometry
MEC	minimum effective concentration
MGG	May-Grünwald Giemsa
MIC	minimum inhibitory concentration
ML	maximum likelihood
MLST	multilocus sequence typing

Msg	major surface glycoprotein (*Pneumocystis* antigen)
NAALADase	N-acetylated α-linked acidic dipeptidase (*H. capsulatum* antigen)
NJ	Neighbor-joining
NSA	niger seed agar
NWT	non-wild type
PAS	periodic acid-Schiff
PCM	paracoccidioidomycosis
PCP	pneumonia by *Pneumocystis carinii*
PCR	polymerase chain reaction
PCR-RFLP	PCR-based restriction fragment length polymorphism
PFGE	pulsed field gel electrophoresis
PJP	*Pneumocystis* pneumonia
PMF	peptide mass fingerprint
PS	psoriasis
PV	pityriasis versicolor
RAPD	random amplification of polymorphic DNA
RCA	rolling circle amplification
rDNA	ribosomal DNA genes
RIA	radioimmunoassay
RSA	recombinant synthetic antigen
SADH	secondary alcohol dehydrogenase
SD	seborrheic dermatitis
SDA	Sabouraud dextrose agar
SHP	hypersensitivity pneumonitis
SPS	sodium polyanethol sulfonate
SSKI	saturated solution of potassium iodide
TP	tube precipitin *Coccidioides* antigen
WB	Western blot
WT	wild type

1
Mycological Diagnosis
General Principles

Rossana de Aguiar Cordeiro

Contents

1.1 Introduction

Mycological laboratories have an important role in the complex scenario of understanding the etiology of fungal infections. They provide information not only related to diagnosis, but also to the treatment, prevention, and control of mycosis. Through such technical data, it is also possible to gain insight into the epidemiology of fungal infections, which brings great responsibilities to the lab staff. Therefore, standardized and internationally validated protocols and routines must be followed.

In order to achieve these goals, mycological labs need to be equipped with biosafety items: individual and community barriers must be available (Table 1.1), and personnel must be engaged in continuous biosafety training. Labs pertaining only to dermatological routines need to be designed to reach level 2 biological safety criteria. On the other hand, labs with complex attendance routines, that is, oncological patients, post-transplant patients, and patients suspected of deep-seated infections, need risk 3 biological safety criteria. Additionally, mycologists need continued education to become familiar with new fungal species and modern diagnostic procedures. A general scheme of mycological diagnosis stages is shown in Figure 1.1.

1.2 Collection of clinical specimens

After clinical examination of the patient by physicians, the first step of mycological diagnosis is to collect clinical specimens. Mycologists should be aware that: (a) specimens must always be collected from the representative site of infection, (b) collection should preferably be performed prior to the start of antifungal therapy, (c) the amount of material collected should be suitable for performing all the required tests, (d) specimens should be transported to the lab and processed as soon as possible, and (e) all clinical material can be infected with other hazardous

Table 1.1   Minimum Biosafety Items Required for Clinical Mycological Laboratories

Primary Barriers		Secondary Barriers	
Individual Protection Equipment	Structural Issues	Inside the Lab	Outside the Lab
Long-sleeved laboratory coats	Restricted containment areas	Sinks for hand washing	Self-closing and lockable doors
Gowns	Class II biological safety cabinet	Bench tops impervious to water and chemicals	Eyewash station
Gloves	Incinerators		Safety shower
Masks and respiratory apparatus	Bioaerosol-contention equipment	Autoclave	Fire extinguisher/ blanket
Eye protection/safety goggles	Containers for disposal of sharp materials	Sealed windows or windows fitted with screens	
Facial protection (face shields/splatter guard)	Mechanical pipetting device		
Shoe covers	Washable floor		
	Sturdy furniture		

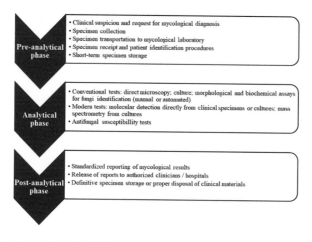

Figure 1.1   Stages of mycological diagnosis in clinical settings.

microorganisms besides fungi. In addition, lab members are responsible for precise identification (patient's full name, ID numbers, hospital identification, etc.) of all specimens collected.

Previous to specimen collection, it is necessary to clean the lesions with 70% isopropanol/ ethanol and then dry the areas with sterile gauze by gentle compressing. Inflammatory/mucosal lesions should be cleaned with sterile water/saline. This procedure considerably reduces the likelihood of cultivation of non-pathogenic species and also removes ointments and dirt from the patient's skin.

Different devices should be available for specimen collection: swab transport systems, screw-cap tubes, needles and syringes, scalpels, tweezers and scissors, surgical blades, microscope slides, etc. (Table 1.2). Clinical specimens should always be collected with sterile devices and transported to the laboratory inside resistant containers as soon as possible. Laboratories should

Table 1.2   Pre-Analytical Phase: General Procedures to Proper Specimen Collection for Fungal Diagnosis

Site	Specimen	Device	Transport Time/ Temperature	Stability/Temperature
Superficial/ mucocutaneous mycoses	Skin	Scalpel Surgical blade Microscopic slide Adhesive tape Swab[a] (inflammatory/intertriginous lesions)	Up to 24 hours/ RT (samples must be protected from moisture)	Up to 4 weeks/RT: skin and nail scrapings, hair Up to 24 hours/RT: swabs
	Nails	Curette Surgical blade		
	Hair	Tweezers Scissors Swab[a] (inflammatory lesions)		
	Mucosa (oral, genital)	Swab[a] Scalpel	Up to 6 hours/RT	Up to 24 hours/RT
	Wounds/abscess	Needle and syringe (exudate) Swab[a]	Up to 2 hours/RT	Up to 24 hours/RT
	Eye	Swab[a] (pre-moistened with sterile saline) for conjunctiva 21-guage needle or Kimura spatula for cornea[c] Microscopic slides and culture media tubes (direct processing by clinician)	Immediately/RT[d]	Up to 12 hours/RT
	Ear	Swab[a] (external otitis; internal otitis with ruptured drum) Syringe (internal otitis with intact drum)[c]	Up to 2 hours/RT	Up to 24 hours/RT
Subcutaneous mycoses	Subcutaneous tissues Abscess/ulcer Skin/mucocutaneous lesions	Punch biopsy[c] and Sterile screw-cap tube/cup with 2 mL sterile saline Needle and syringe (nodular lesions)	Up to 2 hours/RT	Up to 24 hours/RT

(Continued)

Table 1.2 (Continued) Pre-Analytical Phase: General Procedures to Proper Specimen Collection for Fungal Diagnosis

Site	Specimen	Device	Transport Time/Temperature	Stability/Temperature
Systemic mycoses	Abscess/ulcer	Punch biopsy[c] Sterile screw-cap tube/cup Needle and syringe (exudate)	Up to 2 hours/RT	Up to 24 hours/RT
	Blood	Blood culture bottles (aerobe): 5–10 mL for adults; 0.5–5 mL for children[b] Blood collection tubes with EDTA or SPS Microscopic slides	Up to 2 hours/RT	Up to 6 hours/RT[b] Up to 24 hours/RT (non-automated procedure)
	Catheter	Surgical blade Scissors Sterile screw-cap tube/cup	Immediately/RT	Up to 24 hours/4°C
	Cerebrospinal fluid	Sterile screw-cap tube/cup (specimen volume: minimum 2 mL)[c]	Immediately/RT	Up to 24 hours/35°C
	Bone marrow	Sterile screw-cap tube (specimen volume: 1–3 mL)[c] Blood collection tubes with EDTA or SPS Microscopic slides	Immediately/RT	Up to 24 hours/RT
	Pathological fluids (abdominal, pleural, synovial, etc.—aspirated or drained)	Needle and syringe and sterile screw-cap tube (large volumes)[c] Note: anticoagulants are necessary for hemorrhagic specimens.	Immediately/RT	Up to 24 hours/35°C
	Sputum (expectorated/induced)	Sterile, leak-proof, screw-cap container	Immediately/RT	Up to 24 hours/4°C
	Bronchoalveolar lavage; bronchial wash/brush; tracheal aspiration	Sterile, leak-proof, screw-cap container[c]	Immediately/RT	Up to 24 hours/4°C
	Urine (midstream)	Sterile, leak-proof, screw-cap container (specimen volume: minimum 1 mL)	Immediately/RT	Up to 2 hours/4°C (without preservative) Up to 24 hours/4°C (with preservative)

Abbreviations: EDTA, ethylenediamine tetraacetic acid; RT, room temperature (25°C–28°C); SPS, sodium polyanethol sulfonate.
a Swabs must be coupled with a suitable swab transport system.
b Automated procedure: check manufacturer guidelines.
c Procedure exclusively performed by clinicians or medical staff.
d Smears should be prepared by applying scrapings to a glass slide in a gentle circular motion; scrapings should also be inoculated directly on culture media. Both procedures should be performed by the physician immediately after specimen collection.

define exclusion criteria for clinical samples, such as tissues fixed in formalin, dry swabs, leaking/broken containers, etc.

1.3 Mycological processing (conventional tests)

1.3.1 Direct microscopic examination

After arrival at the lab, clinical samples are prepared for microscopic examination. Depending on the material, it may be necessary to perform centrifugation, liquefaction, homogenization, chopping, and/or staining. Many experts consider microscopic examination of clinical samples one of the most important steps of the analytical phase of diagnosis, as it allows quick presumptive results, which may directly impact patient care. However, the success of this step depends on many issues, such as the biological characteristics of the pathogen (i.e., intracellular or extracellular lifestyle), the number of microorganisms in the analyzed sample, and the expertise of the mycologist.

The choice of staining depends on each clinical specimen, but in general, visualization of fungal structures is possible without staining, mainly in clear samples such as skin scrapings, urine, respiratory secretions, and cerebrospinal fluid. Unstained wet mounts can be prepared directly with saline or KOH. Smears must be prepared with a thin layer of clinical specimen. Air-dried smears need to be heat- or alcohol-fixed for proper staining. Table 1.3 describes the general procedures and the presumptive diagnosis regarding clinical samples analyzed by microscopy.

1.3.2 Culture

Cultivation of a clinical specimen allows definitive diagnosis of the infectious agent (except for *Pneumocystis carinii* and *Laccazia loboi*). Cultivation also allows antifungal susceptibility testing. Many fungal species are able to grow on routine bacteriological media (i.e., blood agar, chocolate agar, MacConkey agar), and the relevance of such findings must be analyzed from a clinical perspective. Nevertheless, for selective isolation of fungi from clinical specimens, it is necessary to employ culture media that support fungal growth and suppress bacteria and non-pathogenic fungal species. Sabouraud agar with or without chloramphenicol and cycloheximide compose the tripod for fungal isolation in most of the routine mycological labs. It is important to note that there are differences in cycloheximide tolerance against fungi species. In general, pathogenic fungi can be classified as "very sensitive," "moderately sensitive," and "resistant" to this antibiotic (Table 1.4). Suggestions of culture media for laboratory processing of different clinical samples are shown in Table 1.5. Although there are many advantages of using Petri dishes rather than tubes (larger surface area for isolation of cultures, easier examination and subculture, easier separation of mixed cultures, etc.), one should bear in mind that Petri dishes are not recommended because of biosafety issues, especially if *Coccidioides* or *Histoplasma* is suspected in clinical samples. Although recognized as the gold-standard procedure for fungal identification, use of cultures has limitations: cultures can take many days to show positive growth, and overall, samples from deep infections, including blood cultures, can show moderate recovery rates. Table 1.6 describes formulas of common mycological media and general indications of use.

1.4 Identification

1.4.1 Standard culture-based methods: Phenotypical analysis

Although parasitic structures seen by direct examination and primary culture findings may be suggestive of some species, most of the time it is necessary to perform further biochemical and physiological tests to achieve proper fungal identification. Traditional identification tests rely on morphological analysis of conidia/spores and filamentous structures, combined with carbohydrate and nitrogen assimilation and carbohydrate fermentation tests. Pigment production, enzymatic activity, morphological dimorphism, thermotolerance, and lipodependence are also employed for phenotypical identification of pathogenic fungi. Tests can be performed manually—by cultivation of fungal isolates individually in proper culture media—or by automated or

Table 1.3   Analytical Phase: General Procedures for Direct Microscopic Analysis of Clinical Specimens and Most Common Microscopic Findings

Site	Sample	Procedure and Basic Principle	Most common Findings/Preferred Method	Presumptive Diagnosis
Superficial/mucocutaneous mycoses	**Keratinous (skin, hair, nails) and tissue samples** Material should be reduced to small fragments. Microscopic observation should be performed after 30 minutes of chemical digestion (minimum).	**KOH (10%–30%)** KOH slowly digests and clears the tissues, allowing fungi visualization.	Hyaline, septate hyphae that break up into rectangular arthroconidia (skin)/KOH.	Dermatophytes: *Microsporum*, *Trichosporum*, *Epidermophyton*
		KOH (20%)–DMSO (40%) DMSO causes rapid keratolysis, allowing immediate analysis.	Round blastoconidia; hyaline, septate, thin hyphae, and/or pseudohyphae (skin, nail, and mucosae)/KOH.	*Candida* spp. Note: only *C. albicans* and *C. dubliniensis* produce true hyphae.
		KOH (10%) with Calcofluor white Calcofluor white is a non-specific fluorochrome dye that binds to cellulose and chitin present in the fungal cell wall and septa. Requires a fluorescent microscope equipped with filters bellow 400 nm (ultraviolet) excitation filter. All fungi, including *Pneumocystis* cysts, display a brilliant apple-green/blue-green fluorescence when viewed under ultraviolet or blue light.	Hyaline, septate hyphae, blastoconidia, and rounded arthroconidia (nail).	*Trichosporon* spp.
			White, cream-colored, or light-brown irregular nodules formed by yeast cells and hyphae; nodules easily detached from the hair shaft (scalp, beard, moustache, axillae, genitals, eyebrows, and eyelashes)/KOH.	
			Hyaline, septate hyphae with blunt ends and budding blastoconidia similar to "spaghetti and meatballs" (skin and nail)/KOH.	*Malassezia* spp.
			Hyaline, septate, and tortuous hyphae with acute angle branching; sometimes chlamydoconidia are seen (superficial: nail; mucocutaneous: nasal mucosa, paranasal sinuses, eye)/KOH.	Hyalohyphomycetes *Aspergillus* spp., *Fusarium* spp.
			Dark nodules firmly attached to hair sheath; nodules are formed by septate brown hyphae and round asci containing 2 to 8 hyaline fusiform ascospores/KOH.	*Piedraia hortae*
			Brown-pigmented, septate, tortuous hyphae (skin)/KOH.	*Hortaea werneckii*

(Continued)

Table 1.3 (Continued)   Analytical Phase: General Procedures for Direct Microscopic Analysis of Clinical Specimens and Most Common Microscopic Findings

Site	Sample	Procedure and Basic Principle	Most Common Findings/Preferred Method	Presumptive Diagnosis
Subcutaneous mycoses	**Biopsies** Samples should be reduced to small fragments; imprinting or impression smears could also be performed.	**KOH (10%–30%)** **Grocott-Gomori Methenamine silver stain** Carbohydrates in the fungi cell wall are oxidized and then release aldehyde groups, which react with silver nitrate, reducing it to metallic silver. All fungal structures are seen in brown-black color on a green background.	Thick-walled, multiseptate round cells with brown color ("muriform cells")/KOH. Hyaline, coenocytic, or pauciseptate, ribbon-like, large hyphae with acute angle branching/HE/Grocott-Gomori silver stain/PAS/Calcofluor. Brown-pigmented, septate hyphae/KOH/Fontana Masson.	*Fonsecaea* spp., *Cladosporium* spp., *Rhinocladiella* spp., *Exophiala* spp. Zygomycetes/mucorales: *Mucor* spp., *Rhizopus* spp., *Rhizomucor* spp. *Cladosporium* spp., *Alternaria* spp., *Curvularia* spp., *Rhinocladiella* spp., *Exophiala* spp., *Phialophora* spp.
	Papules, nodules, abscesses, ulcers Samples should be reduced to small fragments; imprinting or impression smears could also be performed.	**Gram** Yeast cells allow crystal violet to penetrate the cell; decolorizer cannot remove the dye and cells appear blue/purple. Filamentous fungi have thicker cell walls that block crystal violet penetration; decolorizer can damage the cell wall and then safranine diffuse inside some fungal cells. Hyphae may be stained pale-pink. **Hematoxylin and eosin—HE (histology)** The basic dye hematoxylin colors the basophilic cellular nucleic acids in a blue-purple hue; eosin colors intracellular and extracellular proteins bright pink. **Calcofluor white-Evans blue** Evans blue is a counterstain that reduces background fluorescence of cells and tissues under blue light (not UV).	Yellow, white, brown, or red grain diameter formed by delicate, branched filaments measuring nearly 5 μm and long chains of spores/KOH/Gram stain. Black, pale, or yellow grains seen as a mass of septate hyphae of nearly 5 μm thick surrounded by intercellular cement/KOH/Gram stain. Small, cigar-shaped, and budding yeast cells (2 to 6 μm in diameter); no filaments are observed/KOH/HE/Grocott-Gomori silver stain/PAS. Globose or lemon-shaped yeast cells in chains connected each other by tubules (isthmus); no filaments are observed/ Grocott-Gomori silver stain.	*Actinomadura madurae,* *A. pelletierii, Nocardia* spp., *Streptomyces* spp. Black grains: *Madurella* spp., *Leptosphaeria* spp., *Curvularia* spp., *Exophiala* spp., *Phialophora* spp. Pale, white, yellow grains: *Pseudallescheria boydii,* *Acremonium* spp., *Aspergillus* spp. *Sporothrix* spp. Notes 1. The number of yeast cells is scarce in the majority of cases; false negative results are common. 2. The Splendore-Hoeppli phenomenon may be seen in histological sections stained by Hematoxylin and eosin. *Lacazia loboi*

Table 1.3 (Continued)   Analytical Phase: General Procedures for Direct Microscopic Analysis of Clinical Specimens and Most Common Microscopic Findings

Site	Sample	Procedure and Basic Principle	Most Common Findings/Preferred Method	Presumptive Diagnosis
Systemic mycoses	**Biopsies** Samples should be reduced to small fragments; imprinting or impression smears could also be performed.	**KOH (10%–30%)** **Gram**	Budding round to oval blastoconidia; thin filaments (hyphae and/or pseudohyphae) may be present/KOH/HE/Grocott-Gomori silver stain/PAS.	*Candida* spp.
		Wright-Giemsa Acidic-nuclear components are turned blue; basic components of the cells are seen orange to pink.	Small intracytoplasmic round to ovoid cells with a small light halo around them ("false capsule")/HE/PAS.	*Histoplasma capsulatum*
		Hematoxylin and eosin (HE) **Grocott-Gomori Methenamine silver stain**	Round to globose cells with multiple synchronous buddings ("Mickey Mouse" or "pilot's wheel" form). Catenulate yeasts may be seen; hyphae are absent/KOH/HE/Grocott-Gomori silver stain.	*Paracoccidioides* spp.
		Calcofluor white-Evans blue **Fontana-Masson (histology)** Melanin reduces ionic silver to metallic silver in alkaline solution; gold chloride tones the metallic silver from light brown to black. A control slide should be run through a depigmentation test before silver impregnation.	Hyaline, septate hyphae with acute angle branching (45°); conidial heads may be seen in pulmonary cavitary lesions/KOH/HE/Grocott-Gomori silver stain.	*Aspergillus* spp.
			Round to oval refractile cells, yeast may be seen inside giant cells or extracellularly. Narrow budding yeasts with teardrop format surrounded or not by a clear halo/Fontana Masson/PAS/Grocott-Gomori silver stain.	*Cryptococcus neoformans* *C. gattii*
		Periodic acid-Schiff—PAS (histology) Periodic acid oxidizes carbohydrates with hydroxyl group or amino/alkylamine group resulting in aldehydes that react with Schiff's reagent, forming an insoluble magenta to pink colored complex.	Large spherules filled with multiple endospores; ruptured spherules/Grocott-Gomori silver stain/PAS/HE.	*Coccidioides immitis* *Coccidioides posadasii*

(Continued)

Table 1.3 (Continued)   Analytical Phase: General Procedures for Direct Microscopic Analysis of Clinical Specimens and Most Common Microscopic Findings

Site	Sample	Procedure and Basic Principle	Most Common Findings/Preferred Method	Presumptive Diagnosis
	Blood bone marrow, blood, buffy coat Prepare at least two smears for each sample.	**Wright-Giemsa**	Intracellular (macrophages/neutrophils) round or oval yeasts with a small light halo around them ("false capsule"). Fungal cytoplasm is seen as light stained blue, and the polar nucleus retains the dye intensely.	*Histoplasma capsulatum*
			Budding round to oval blastoconidia; thin filaments (hyphae and/or pseudohyphae) branched or not may be present.	*Candida* spp. Note: *Candida* is rarely seen in blood smears.
	Cerebrospinal fluid Samples should be concentrated by centrifugation; wet mounts and smears are prepared with the obtained pellet.	**India ink/Nigrosin** Negative staining: live cells and capsular cells do not allow dye diffusion.	Tear-shaped or round cells with narrow-based buds surrounded by a large bright halo (capsule) with a black background/India ink.	*Cryptococcus neoformans C. gattii.*
		Gram	Round to oval budding blastoconidia/Gram.	*Candida* spp.
	Pathological fluids (abdominal, pleural, synovial, etc.—aspirated or drained) Samples must be concentrated by centrifugation; wet mounts and smears are prepared with the obtained pellet.	**Gram** **Grocott-Gomori Methenamine silver stain** **Calcofluor White stain** **Wright-Giemsa**	Budding round to oval blastoconidia associated or not with thin septate filaments (hyphae and/or pseudohyphae)/Gram/ Calcofluor/Grocott-Gomori silver stain.	*Candida* spp.
	Sputum (expectorated/induced) Samples should be mixed previously with n-acetylcysteine or dithiothreitol for liquefaction. Samples should be concentrated by centrifugation.	**Gram** **KOH (10%–30%)** **Grocott-Gomori Methenamine silver stain** **Calcofluor plus KOH** **Wright-Giemsa**	Large yeast-like cells (up to 30 μm) round to oval; multiple buds are seen in a "pilot wheel" configuration /KOH/Grocott-Gomori silver stain/Calcofluor.	*Paracoccidioides* spp.
			Large spherules filled with multiple endospores; ruptured spherules/KOH/ Grocott-Gomori silver stain/Calcofluor.	*Coccidioides immitis Coccidioides posadasii*

(Continued)

Table 1.3 (Continued) Analytical Phase: General Procedures for Direct Microscopic Analysis of Clinical Specimens and Most Common Microscopic Findings

Site	Sample	Procedure and Basic Principle	Most Common Findings/Preferred Method	Presumptive Diagnosis
	Bronchoalveolar lavage; bronchial wash/brush; tracheal aspiration Samples should be concentrated by centrifugation.		Thin-walled spherules or thin-walled cells with "deflated ball" shape; collapsed crescent-shaped cysts and cysts with dark staining areas/Grocott-Gomori silver stain/Giemsa.	*Pneumocystis jiroveci*
			Hyaline, septate hyphae with branching at 45°/KOH/Grocott-Gomori silver stain/Calcofluor.	*Aspergillus* spp.
			Narrow-budding yeast cells; variably sized oval or round cells; bright halo (capsule) may be seen/KOH/Grocott-Gomori silver stain.	*Cryptococcus neoformans* *C. gattii*
Systemic mycosis	**Urine** (midstream) **KOH (10%–30%)** If large debris are observed in the pellet. Samples should be concentrated by centrifugation; wet mounts and smears are prepared with the obtained pellet.	**Gram**	Budding round to oval blastoconidia associated or not with thin septate filaments (hyphae and/or pseudohyphae)/Gram/KOH.	*Candida* spp.
	Skin/mucocutaneous lesions (after dissemination)	**KOH (10%–30%)** **Grocott-Gomori Methenamine silver stain** **Periodic acid-Schiff—PAS (histology)** **Fontana Masson**	See descriptions above.	*Histoplasma capsulatum* *Cryptococcus neoformans* *C. gattii* *Paracoccidioides* spp. *Coccidioides immitis* *Coccidioides posadasii*

Table 1.4   Variation in Sensitivity to Cycloheximide among Fungi Species

Very Sensitive 0.1 mM (0.02 mg/mL)	Moderately Sensitive 0.18–1.8 mM (0.05–0.5 mg/mL)	Resistant >1.8 mM (>0.5 mg/mL)
Aspergillus spp. (including A. nidulans)	Candida parapsilosis[a]	Candida albicans, C. dubliniensis
Penicillium spp.	Candida tropicalis[a]	Trichophyton spp.
Scytalidium hyalinum/S. dimidiatum	Candida glabrata[a]	Microsporum spp.
Cryptococcus neoformans	Scopulariopsis spp.	Epidermophyton floccosum
Cryptococcus gattii	Cladosporium spp.	Sporothrix schenckii species complex
Candida krusei	Scedosporium apiospermum complex	Histoplasma capsulatum
Candida utilis	Trichosporon cutaneum	Coccidioides immitis/C. posadasii
Candida auris	Trichosporon inkin	Paracoccidioides brasiliensis species complex
Rhodotorula glutinis	Trichosporon mucoides	Fusarium solani species complex
Trichosporon ovoides	Zygomycetes	Fusarium fujikuroi species complex
		Acremonium spp.
		Scopulariopsis spp.
		Exophiala dermatitidis
		Fonsecaeae pedrosoi
		Hortaea werneckii

[a] Variable sensitivity among strains.

semi-automated systems (API 20 C AUX, bioMérieux; Vitek, bioMérieux; RapID Yeast Plus system, ThermoFisher; MicroScan rapid yeast identification panel, Siemens Healthcare; etc.). Most such platforms, however, are designed only for yeast identification. Chromogenic media are also widely used in clinical labs for presumptive identification of Candida spp., being especially useful for visualization of mixed cultures (yeast/bacteria contamination).

1.4.2 Culture-independent methods

Presumptive diagnosis of fungal infections can also be achieved by molecular or immunological methods.

1.4.2.1 Molecular diagnosis – Diagnosis based on molecular techniques has as potential applications the identification of pathogens in non-cultivable materials (such as histological specimens already fixed), as well as the detection of microorganisms directly in clinical specimens, contaminated clinical specimens, and mixed cultures. Molecular techniques are also important alternatives to the conventional diagnosis of high-risk pathogens, such as Coccidioides spp. and Histoplasma capsulatum, since they can eliminate the handling of infectious cultures and allow the detection of fungal DNA before seroconversion.

The extraction of nucleic acids from fungal pathogens in clinical samples can be performed with commercial kits that allow obtaining fungal DNA and/or RNA after lysis of the sample and adsorption of nucleic acid in affinity columns. These kits involve easy-to-manipulate protocols with quick results, but their use is limited to only a few clinical samples, such as blood, serum, or biopsies. This issue can be problematic, especially for the diagnosis of some deep mycoses for which respiratory manifestations are the main findings and fungemia is rare. In this way, the viability of molecular techniques using blood as the only biological sample is very questionable. In addition, lung biopsies are difficult to obtain and are not considered preferential samples for the diagnosis of deep mycoses. In laboratory routine, it is important to choose a method also able to test fungal cultures.

In order to ensure the precise identification of a given microorganism, it is necessary to determine the genetic regions that can be used as a diagnostic target. Molecular tests based on genes

Table 1.5   Suggestion of Culture Media for Laboratory Processing of Different Clinical Samples and More Common Fungal Species Isolated

Clinical Samples	Suggested Media	Most Likely Isolates
Keratinous samples (skin, hair, nail, ear)	Sabouraud dextrose agar (SDA), SDA with chloramphenicol (SDAc), SDA with choramphenicol and cycloheximide (SDAcc) Dixon or Sabouraud dextrose agar with olive oil for *Malassezia* spp.	Skin: *Candida, Trichosporon,* dermatophytes, *Malassezia* Hair: *Trichosporon, Piedra,* dermatophytes, *Malassezia* Nail: *Candida, Trichophyton, Acremonium, Aspergillus, Fusarium, Scopulariosis, Scytalidium, Malassezia* Ear: *Aspergillus, Candida, Malassezia*
Swabs (oral cavity, nose, eye, vagina, urethra, wounds, ulcer, external ear)	SDA, SDAc, SDAcc	*Aspergillus, Candida, Fusarium, Rhizopus*
Blood	Aerobic blood culture bottles	*Candida, Cryptococcus, Trichosporon, Histoplasma* (filamentous fungi are rarely isolated)
Sputum and respiratory samples	SDA, SDAc, SDAcc, BHI agar (brain heart infusion agar)	*Candida, Cryptococcus, Coccidioides, Histoplasma, Aspergillus, Paracoccidioides, Mucor*
Liquor	SDA, SDAc, SDAcc	*Candida, Cryptococcus, Coccidioides, Histoplasma*
Tissue, biopsy	SDA, SDAc, SDAcc, BHI agar	Subcutaneous and systemic pathogens
Urine and other body fluids (except liquor)	SDA, SDAc, SDAcc, BHI agar	Urine: *Candida, Cryptococcus* Chest, abdominal, and synovial fluids: *Aspergillus, Candida* Bone marrow: *Histoplasma, Candida, Cryptococcus*
Pus and other exudates	SDA, SDAc, SDAcc, BHI agar	Subcutaneous and systemic pathogens plus *Trichophyton schoenleinii*
Feces	SDA, SDAc, SDAcc	*Candida* spp.

present in multiple copies per genome—such as the cluster of ribosomal DNA genes (rDNA)—show high sensitivity but may present moderate specificity. Genes for fungal rDNA are arranged as repeating units formed by different informative sequences and therefore are considered relevant molecular markers for identification of many fungi species (Figure 1.2). Molecular identification can also be targeted to genetic sequences present in single copies per genome. Tests with this approach may have less sensitivity, but generate very specific results, of great importance in laboratory diagnosis. Examples of non-rDNA loci used for fungal identification are: translation elongation factor 1α (for *Fusarium, Trichoderma, Mucorales,* etc.), cytochrome oxidase 1 (for *Penicillium* spp.), calmodulin (for *Aspergillus* spp., *Sporothrix* spp.), yeast-phase genes RYP (for *Histoplama capsulatum*), etc.

Among the various molecular techniques employed in mycology, polymerase chain reaction is probably the most applicable in clinical laboratories. Molecular identification protocols based on this technique generally have high sensitivity, low cost, and reproducible results.

Throughout this work, the reader will be presented with the most common techniques for the molecular diagnosis of pathogens generally found in mycology laboratories.

Table 1.6   Descriptive Formulas of Common Mycological Media and Indications of Use

Medium	Formula	Indications of Use
BHI	Brain heart (infusion) 10 g/L; Pancreatic digest of casein 16 g/L; Peptic digest of animal tissue 5 g/L, Sodium chloride 3.0 g/L; Disodium phosphate 2.5 g/L; Dextrose 2.0 g/L; Cycloheximide 0.5 g/L; Chloramphenicol 0.05 g/L; Agar 13.5 g/L Final pH (25°C) 7.4 ± 0.2	Primary recovery of fastidious pathogenic fungi from clinical samples.
Bird seed agar (Niger seed agar/ Staib agar)	*Guizotia abyssinica* seeds 50 g/L; Creatinine 1 g/L; Dextrose 1 g/L; Monopotassium phosphate 1.0 g/L; Chloramphenicol 0.05 g/L; Agar 15 g/L Final pH (25°C) 6.7 ± 0.2	Subculture of suggestive colonies of *Cryptococcus*. Selective isolation and differentiation of *Cryptococcus neoformans* from other yeasts, including other *Cryptococcus* species.
Caffeic acid agar	Ammonium sulfate 5 g/L; Dextrose 5 g/L; Yeast extract 2 g/L; Dipotassium phosphate 0.8 g/L; Magnesium sulfate 0.7 g/L; Caffeic acid 0.18 g/L; Chloramphenicol 0.05 g/L; Ferric citrate 0.02 g/L; Agar 15 g/L Final pH (25°C) 6.5 ± 0.2	Subculture of suggestive colonies of *Cryptococcus*. For the selective isolation and differentiation of *Cryptococcus neoformans* and *C. gattii* from other yeasts.
CHROMagar Candida	Peptone 10.2 g/L; Chromogenic mix 22 g/L; Chloramphenicol 0.5 g/L; Agar 15 g/L Final pH (25°C) pH: 6.1 ± 0.2	Subculture of suggestive colonies of *Candida*. Selective and presumptive differential identification of *C. albicans, C. tropicalis, C. glabrata, C. krusei*.
CGB agar	Glycine 10 g/L; Potassium phosphate 1 g/L; Magnesium sulfate 1 g/L; Bromothymol blue 0.4 g/L; L-Canavanine sulfate 30 mg/L; Thiamine HCl 1 mg/L; Agar 15 g/L Final pH (at 25°C) 5.8 ± 0.1	Subculture of suggestive colonies of *Cryptococcus*. Selective and presumptive differential identification of *Cryptococcus gattii* from other *Cryptococcus* spp. *C. gattii* can grow in the presence of L-canavanine and utilize glycine as sole carbon source; bromothymol is an indicator that changes the color of the medium from yellow–green to cobalt blue in alkaline pH.
Corn meal agar with Tween 80	Cornmeal (infusion) 50 g/L; Tween 80 10 mL/L; Agar 15 g/L Final pH (25°C) 6.0 ± 0.2	For demonstration of chlamydospore production by *Candida albicans*.

(Continued)

Table 1.6 (*Continued*)   Descriptive Formulas of Common Mycological Media and Indications of Use

Medium	Formula	Indications of Use
Dermatophyte test agar	Papaic digest of soyabean meal 10 g/L; Dextrose 10 g/L; Phenol red 0.2 g/L; Agar 2 g/L; Cycloheximide 0.5 g/L; Gentamicin 0.1 g/L; Chloramphenicol 0.1 g/L Final pH (25°C) 5.5 ± 0.2	Primary recovery of dermatophytes from clinical samples. Dermatophytes appear as pink-red; saprophyte fungi (non-dermatophytes) can be recognized by the absence of color change from yellow to red.
Dixon agar	Malt extract 30 g/L; Ox bile 20 g/L; Mycological peptone 5 g/L; Glycerol mono-oleate 2.5 g/L; Tween 40 10 mL; Agar 15 g/L Final pH (25°C) 5.4 ± 0.2	Primary recovery of *Malassezia* spp. from clinical samples.
Modified Czapek-Dox	Sucrose 30 g/L; Sodium nitrate 2 g/L; Magnesium glycerophosphate 0.5 g/L; Potassium chloride 0.5 g/L; Dipotassium sulfate 0.35 g/L; Ferrous sulfate 0.01 g/L; Agar 12 g/L Final pH (25°C) 6.8 ± 0.2	Induction of chlamydospore production by *Candida albicans*. Subculture of *Aspergillus* spp. and *Penicillium* spp.
SABHI agar	Beef heart (infusion) 125 g/L; Calf brain, (infusion) 100 g/L; Dextrose 21 g/L; Neopeptone 5 g/L; Proteose peptone 5 g/L; Sodium chloride 2.5 g/L; Disodium phosphate 1.25 g/L; Chloramphenicol 0.05 mg/L or Gentamycin 40 mg/L; Cycloheximide 500 mg/L; Agar 15 g/L. It is recommended to add 10% sterile sheep blood before dispensing into Petri dishes. Final pH (25°C) 7.0 ± 0.2	Primary recovery of pathogenic fungi from clinical samples. Blood enhances the recovery of fastidious fungi and also the in vitro conversion of *Histoplasma capsulatum* to the yeast phase.
Sabouraud dextrose agar 2%	Dextrose 20 g/L; Peptone 10 g/L; (Pancreatic digest of casein 5 g/L; Peptic digest of animal tissue 5 g/L); Agar 15 g/L Final pH (25°C) 5.6 ± 0.2 Addition of chloramphenicol 0.05 g/L and cycloheximide 0.5 g/L suppress the growth of bacteria and saprophytic fungi.	Primary recovery of pathogenic fungi from clinical samples.
Sabouraud dextrose agar with olive oil	Dextrose 40 g/L; Pancreatic digest of casein 5 g/L; Peptic digest of animal tissue 5 g/L; Olive oil 20 mL, Tween 80 2 mL; Agar 15 g/L Final pH (25°C) 5.6 ± 0.2	Primary recovery of *Malassezia* spp. from clinical samples.

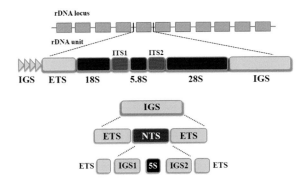

Figure 1.2   General structure and organization of ribosomal DNA genes. DNA unit repeats are formed by a sequence encoding 18S rDNA (for the small ribosomal subunit), internal transcribed spacers 1 and 2 (ITS1 and ITS2), nucleotide sequences encoding 5.8S rDNA and 28S rDNA (for the large ribosomal subunit), and intergenic spacer (IGS). The latter is composed of external transcribed spacers (ETS) and non-transcribed spacers (NTS), which, in turn, are formed by a sequence encoding the 5S rDNA flanked by IGS regions. Variable numbers of subrepeats are found in the IGS region.

1.4.2.2 Immunological diagnosis – Immunological tests are important for the presumptive diagnosis of some fungal infections, as they do not require invasive procedures to obtain clinical specimens and circumvent the need for biosafety conditions for manipulation of high-risk pathogens.

Usually, in systemic fungal infections, antibodies have no protective activity, but serve as tools for the diagnosis, prognosis, and follow-up of patients. Antibody screening can be performed with several clinical specimens, such as serum, cerebrospinal fluid, pleural fluid, peritoneal fluid, synovium, and urine, depending on the standardization of each method and the antigen used. The tests can be performed with samples from immunosuppressed patients, although in some cases it is necessary to combine more than one diagnostic method.

In systemic mycoses, after the onset of symptoms, IgM positivity usually occurs between the first and third weeks, with positivity 6 months after infection being rare. Approximately 90% of individuals with systemic mycoses show reactivity to IgG-directed tests by the fourth week after symptom onset. The detection of IgG is described as positive evidence of current or recent infection, although antibodies can be detected in some individuals for more than 1 year after clinical cure.

Even after chemical purification, commercial antigens used in the immunodiagnosis of systemic mycoses may show cross-reactions with other infections. Therefore, the results obtained must always be evaluated through a clinical-epidemiological approach. Negative results, on the other hand, do not exclude the possibility of fungal infection. The most common diagnostic methods for detecting the humoral response of systemic mycoses are gel immunodiffusion, complement fixation, and enzyme immunoassay.

Table 1.7 describes some important commercially available tests for the immunological diagnosis of systemic mycoses.

1.4.2.3 Matrix-assisted laser desorption ionization-time of flight mass spectrometry – In recent years, matrix-assisted laser desorption ionization-time of flight mass spectrometry (MALDI-TOF MS) has emerged as a powerful technology for fungal identification. During the MALDI-TOF MS process, primary cultures are resuspended in an energy-absorbing chemical matrix and either intact cells or cell extracts are ionized by a laser beam. The resulting ions migrate through a charged field in a vacuum tube and are detected by a mass analyzer. This generates a spectrum called a "peptide mass fingerprint" for the sample, which is then

Table 1.7 Commercial Immunological Tests Used in the Diagnosis of Fungal Infections

Fungal Species	Clinical Sample	Target	Method	Commercial Kits
Aspergillus fumigatus, A. flavus, A. niger, A. terreus	Serum BAL	Galactomannan antigen	Enzyme immunoassay	Platelia Aspergillus EIA (Bio-Rad Laboratories, UK)
	Serum	IgG	Lateral flow	sōna Aspergillus Galactomannan LFA (IMMY Immuno-Mycologics Inc., USA)
Cryptococcus neoformans C. gattii	BAL	Glucuronoxylomannan antigen	Lateral flow	CrAg LFA (IMMY Immuno-Mycologics Inc., USA)
	CSF		Latex agglutination	CALAS (Meridian Bioscience Inc., USA)
				Murex Cryptococcus Test (Remel, USA)
				Crypto-LA test (Wampole Laboratories, USA)
				Cryptococcus Antigen Latex Agglutination Test System (IMMY Immuno-Mycologics Inc., USA)
			Enzyme immunoassay	PREMIER Cryptococcal EIA Antigen assay (Meridian Bioscience Inc., USA)
				ALPHA Cryptococcal Antigen enzyme immunoassay (IMMY Immuno-Mycologics Inc., USA)
Candida albicans	Serum	Germ tube antibodies (IgG)	Indirect immunofluorescence	Candida albicans IFA IgG (VIRCELL SL, Spain)
	Serum or plasma BAL or CSF (alternatively)	Mannan antigen	Enzyme immunoassay	Platelia Candida Ab EIA kit (Bio-Rad Laboratories, Germany)
	Serum	Antibodies against intracytoplasmatic antigens—predominantly enolase (IgG)	Enzyme immunoassay	SysCan3 (Rockeby Biomed Ltd, Singapore)

(Continued)

Table 1.7 (Continued) Commercial Immunological Tests Used in the Diagnosis of Fungal Infections

Fungal Species	Clinical Sample	Target	Method	Commercial Kits
Coccidioides immitis C. *posadasii*	Serum	IgG/IgM	Lateral flow	sōna *Coccidioides* Ab LFA (IMMY Immuno-Mycologics Inc., USA)
			Immunodiffusion	ID IMMY Immuno-Mycologics Inc., USA
				Coccidioides Immunodiffusion System (Meridian Bioscience Inc., USA)
			Latex agglutination	LA-*Coccidioides* (IMMY Immuno-Mycologics Inc., USA)
			Complement fixation	CF-*Coccidioides* (IMMY Immuno-Mycologics Inc., USA)
			Enzyme immunoassay	*Coccidioides* Antibody EIA (IMMY Immuno-Mycologics Inc., USA)
				Coccidioides Antibody IgG and IgM EIA (MiraVista Diagnostics, EUA)
	Serum, plasma, urine, CSF, BAL, other body fluids	Galactomannan antigen	Enzyme immunoassay	*Coccidioides*-specific Antigen EIA (MiraVista Diagnostics, EUA)
Histoplasma capsulatum	Serum	IgM/IgG	Immunodiffusion	IMMY Immuno-Mycologics Inc., USA
			Complement fixation	CF (IMMY Immuno-Mycologics Inc., USA)
			Latex agglutination	LA-*Histoplasma* (IMMY Immuno-Mycologics Inc., USA)
	Urine	Galactomannan antigen	Enzyme immunoassay	ALPHA *Histoplasma* EIA (IMMY Immuno-Mycologics Inc., USA)
	Serum, plasma, urine, CSF, BAL, other body fluids			*Histoplasma* antigen EIA (MiraVista Diagnosis, USA)
Invasive fungal pathogens (*Candida* spp., *Acremonium* spp., *Aspergillus* spp., *Coccidioides* spp., *Fusarium* spp., *Histoplasma capsulatum*, *Trichosporon* spp., *Sporothrix schenckii*, *Pneumocystis jiroveci*), except invasive zygomycosis, cryptococcosis, and blastomycosis.	Serum BAL and CSF (alternatively)	$(1,3)$-β-D-Glucan antigen	Colorimetric assay	Fungitell (Associates of Cape Cod, USA)
				β-glucan test (Wako Pure Chemical Industries, Japan)

Abbreviations: BAL, Bronchoalveolar lavage; CSF, Cerebrospinal fluid.

compared with a database that contains known spectra. MALDI-TOF MS is a fast, practical, and accurate methodology, and individual analyses are less expensive than molecular and immunological tests. However, the initial cost of MALDI-TOF MS equipment is very high, often prohibitive for inclusion by clinical laboratories. In addition, manufacturer-provided databases may be limited and rare species may not be included. At present, two commercial systems for fungal identification based on MALDI-TOF are provided, one by bioMérieux, Inc. (Durham, NC, USA) and the other by Bruker Daltonics, Inc. (Billerica, MA, USA). Their libraries for identification of yeasts and filamentous fungi include more than 100 species.

1.5 Conclusions

Routine mycological diagnosis is a neglected field, although fungal infections have become prevalent diseases worldwide. They affect billions of people, and it is estimated that over 300 million people in the world suffer from serious fungal infection every year. Modern diagnostic tools necessary for proper diagnosis are usually not available in low-income countries. However, every laboratory needs to provide basic services that allow detection and recovery of fungal species in clinical samples. Such services may have an important impact on therapy as well as on the reduction of morbidity and mortality rates. Collaboration between clinicians and mycologists is necessary to achieve more realistic results.

Bibliography

Ampel, N. M. 2003. Measurement of cellular immunity in human coccidioidomycosis. *Mycopathologia* 156: 247–262.
Badiee, P. 2013. Evaluation of human body fluids for the diagnosis of fungal infections. *Biomed Res Int* 2013: 698325.
Blair, J. E., Coakley, B., Santelli, A. C., Hentz, J. G., Wengenack, N. L. 2006. Serologic testing for symptomatic coccidioidomycosis in immunocompetent and immunosuppressed hosts. *Mycopathologia* 162: 317–324.
Cuenca-Estrella, M., Bassetti, M., Lass-Flörl, C., Rácil, Z., Richardson, M., Rogers, T. R. 2011. Detection and investigation of invasive mould disease. *J Antimicrob Chemother* 66 Suppl 1: 15–24.
de Hoog, G. S., Guarro, J., Gené, J., Figueras, M. J. 2000. *Atlas of Clinical Fungi.* 2nd edition. Utrecht, The Netherlands: Centraalbureau voor Schimmelcultures/Universitat Rovira i Virgili.
Kozel, T. R., Wickes, B. 2014. Fungal diagnostics. *Cold Spring Harb Perspect Med* 4: a019299.
Lau, A., Chen, S., Sleiman, S., Sorrell, T. 2009. Current status and future perspectives on molecular and serological methods in diagnostic mycology. *Future Microbiol* 4: 1185–222.
Marcos, J. Y., Pincus, D. H. 2013. Fungal diagnostics: Review of commercially available methods. *Methods Mol Biol* 968: 25–54.
Miller, J. M., Binnicker, M. J., Campbell, S., Carroll, K. C., Chapin, K. C., Gilligan, P. H. 2018. A guide to utilization of the microbiology laboratory for diagnosis of infectious diseases: 2018 update by the Infectious Diseases Society of America and the American Society for Microbiology. *Clin Infect Dis* 67: e1–e94.
Pappagianis, D. 2001. Serologic studies in coccidioidomycosis. *Semin Respir Infect* 16: 242–250.
Patel, R. 2019. A moldy application of MALDI: MALDI-ToF mass spectrometry for fungal identification. *J Fungi (Basel)* 5: E4.
Procop, G. W., Koneman, E. W. 2017. *Koneman's Color Atlas and Textbook of Diagnostic Microbiology.* 17th edition. Philadelphia: Wolters Kluwer Health.
Walsh, T. J., Hayden, R. T., Larone, D. H. 2018. *Larone's Medically Important Fungi: A Guide to Identification.* 6th edition. Washington, DC: ASM Press.
Wickes, B. L., Wiederhold, N. P. 2018. Molecular diagnostics in medical mycology. *Nat Commun* 9: 5135.

2

Antifungal Drugs and Susceptibility Testing of Fungi

Débora de Souza Colares Maia Castelo-Branco,
Glaucia Morgana de Melo Guedes, and
Marcos Fábio Gadelha Rocha

Contents

2.1 Antifungal drugs

The therapeutic arsenal of antifungal drugs is narrow when compared to that of antibacterial drugs, whose mechanisms of action can be grouped into three cellular sites of action, encompassing eight classes of drugs currently available for clinical use, as described below.

2.1.1 Cell membrane activity

2.1.1.1 Formation of pores on cell membrane
***2.1.1.1.1 Polyene derivatives* –** Since the 1950s, the fungicidal polyenes, such as nystatin and amphotericin B, have been used as a powerful but highly toxic last line of defense against invasive fungal infections. These are amphipathic fungicidal drugs consisting of a hydrophobic polyene hydrocarbon chain and a hydrophilic polyhydroxyl chain. For decades, the prevailing theory was that these molecules directly bound to the ergosterol molecule embedded in the phospholipid bilayer of the fungal cell membrane, creating pores on the plasmatic membrane, leading to leakage of cellular components and death. However, recent biophysical studies highlighted that polyenes act like an "ergosterol-sponge," forming large extramembranous aggregates that extract the essential membrane-lipid ergosterol from the plasma membrane (Robbins et al., 2016). Table 2.1 shows a summary of antifungal drugs and their mechanisms of action and uses. Nystatin is highly toxic and is only available for topical use in the form of cutaneous and/or mucosal creams. Considering it is not absorbed by the gastrointestinal tract, nystatin is also available as an oral suspension to be used as a topical treatment for oral candidiasis. Amphotericin B, on the other hand, is available for systemic use, despite its toxicity, and it is

Table 2.1 Antifungal Drugs and Their Mechanisms of Action, Use, and Antifungal Spectrum

Site of Action	Drug Class	Mechanism of Action	Drugs	Use	Antifungal Spectrum
Cell membrane	Polyenes	Ergosterol binding/extraction creating membrane pores.	Nystatin	Topical or oral	Yeasts, *Candida* spp.
			Amphotericin B	Intravenous	Yeasts
					Filamentous fungi (except for *Aspergillus terreus*)
					Dimorphic fungi
	Azole derivatives	Inhibition of ergosterol synthesis, by blocking the enzyme 14-α-demetilase, and accumulation of toxic compounds.	*Imidazoles*	Topical	Yeasts—*Candida* spp., *Malassezia* spp.
			Miconazole		Dermatophytes
			Ketoconazole	Topical or oral	
			Triazoles		Yeasts (except for *Candida krusei*)
			Fluconazole	Oral or intravenous	Dimorphic fungi (less effective)
			Itraconazole	Oral	Yeasts
			Voriconazole	Oral or intravenous	Filamentous fungi
			Posaconazole	Oral	Dimorphic fungi
	Allylamines	Inhibition of ergosterol synthesis, by blocking squalene monoxygenase, and accumulation of toxic compounds.	Naftifine	Topical	Yeasts—onychomycosis
					Dermatophytes
			Terbinafine	Topical or oral	Yeasts—onychomycosis
					Dermatophytes
					Sporothrix spp.
	Morpholinic derivatives	Inhibition of ergosterol synthesis.	Amorolfine	Topical	Dermatophytes
Intracellular	Griseofulvin	Binding to microtubular proteins, disruption of the spindle apparatus, inhibition of mitosis.	Griseofulvin	Topical or oral	Dermatophytes
	Pyrimidines	Inhibition of the enzyme thymydilate synthase, leading to RNA miscoding and inhibition of DNA synthesis.	Flucytosine	Oral	Yeasts—*Cryptococcus* spp. and *Candida* spp.
	Pyridine derivatives	Chelation of trivalent cations, leading to the inhibition of metal-dependent enzymes.	Ciclopirox	Topical	*Candida* spp.
					Dermatophytes
Cell wall	Echinocandins	Inhibition of synthesis of (1,3)-β-D-glucan, leading to the loss of cell wall integrity.	Caspofungin	Intravenous	*Candida* spp.
			Micafungin		*Aspergillus* spp.
			Anidulafungin		

currently the last-line treatment for systemic fungal infections. Lipid, liposomal, and colloidal formulations of amphotericin B have been developed as an attempt to decrease its toxicity, but they are much more expensive than the regular deoxycholate formulation.

2.1.1.2 Inhibition of ergosterol synthesis

2.1.1.2.1 Azole derivatives – The azole derivatives encompass compounds with both imidazole and triazole moieties. They are the most widely used class of antifungal in clinical practice, due to their safety and availability in both oral and intravenous formulations. These drugs interfere with ergosterol synthesis by blocking the enzyme 14-α-demetilase present in the P-450 cytochrome of the fungal cell, hampering lanosterol demethylation into ergosterol. This interference results in ergosterol depletion and the accumulation of toxic sterol intermediates, which stops cellular growth and division and induces severe membrane stress. Most of the antifungal imidazoles are formulated for topical use, as they are more toxic and less bioavailable, limiting their systemic use. On the other hand, triazoles, which include fluconazole, itraconazole, voriconazole, posaconazole, ravuconazole, and isavuconazole, are licensed for the treatment of invasive fungal diseases. The azoles are mostly fungistatic, with the exception of the fungicidal activity of itraconazole and voriconazole against some species of filamentous fungi, such as *A. fumigatus* (Table 2.1).

2.1.1.2.2 Allylamines – Allylamines, such as naftifine, terbinafine, and the related benzylamine butenafine, inhibit ergosterol biosynthesis by inhibiting squalene monoxygenase, an enzyme in fungi responsible for the conversion of squalene to squalene epoxide, which is a precursor to lanosterol in the ergosterol synthesis pathway. This inhibition causes the intracellular accumulation of squalene, which is toxic to fungal cells, leading to fungal death. Allylamines effectively bind to the stratum corneum because of their lipophilic nature and also penetrate deeply into hair follicles. The only drug of this class available for systemic use is terbinafine, and, once taken orally, it concentrates in the skin and nail beds, presenting relatively low bloodstream concentrations. Consequently, its use as a systemic antifungal agent is primarily restricted to the treatment of onychomycosis and cutaneous fungal infections, especially against dermatophytes (Table 2.1).

2.1.1.2.3 Morpholinic derivatives – The morpholinic derivatives are represented by amorolfine, which also inhibits ergosterol synthesis. *In vitro* studies show that it is active against several types of fungi, including dermatophytes, dimorphic fungi, yeasts, and dematiaceous fungi, but *in vivo* studies show that it is inactive when given systemically, limiting its use as topic formulations, including nail lacquers, creams, and alcohol solutions. It is mainly used to treat onychomycoses, without nail matrix involvement, and dermatomycoses (Table 2.1).

2.1.2 Intracellular activity

2.1.2.1 Griseofulvin – Griseofulvin, a metabolic product of *Penicillium griseofulvum*, prevents fungal growth by binding to microtubular proteins, which leads to the disruption of the spindle apparatus, thus inhibiting fungal cell mitosis. This compound was the first oral agent available for the treatment of dermatomycoses and is only used for noninvasive dermatophyte infections, because the drug concentrates in keratinocytes (Table 2.1).

2.1.2.2 Pyrimidines – Pyrimidines, such as flucytosine (5-fluorocytosine), were first used to treat fungal infections in the 1960s. Flucytosine is a synthetic fluorinated analog of cytosine. When it enters the cytosol, flucytosine is rapidly deaminated by fungal-specific cytosine deaminases, generating 5-fluorouracil, which acts as an antimetabolite by inhibiting the enzyme thymydilate synthase, leading to RNA miscoding and inhibition of DNA synthesis. The fast emergence of resistance hampers the use of flucytosine as a monotherapy; hence, it is only used in association with other antifungals, especially amphotericin B. This drug combination remains the "gold standard" for treating *Cryptococcus* infections; however, flucytosine is of limited availability in developing countries (Table 2.1).

2.1.2.3 Pyridine derivatives – This class of drug is represented by ciclopirox, which exerts its antifungal activity by chelating trivalent cations, including Fe^{3+} and Al^{3+}, which results in the inhibition of metal-dependent enzymes, particularly cytochromes, catalases, and peroxidases, leading to reduced transport of ions through pathogen cytoplasmatic membranes and reduced nutrient intake. It is marketed in many topical fomulations, including nail lacquers, creams, shampoos, and solutions, and it is used for the treatment of skin and scalp mycoses, onychomycoses, pityriasis versicolor, and vulvovaginal candidiasis (Table 2.1).

2.1.3 Cell wall activity

2.1.3.1 Echinocandins – The echinocandins are the only new class of antifungal drugs to reach clinical use in decades, with three currently available drugs: caspofungin, micafungin, and anidulafungin. These compounds are cyclic hexapeptides that act as noncompetitive inhibitors of (1,3)-β-D-glucan synthase, an enzyme involved in the synthesis of (1,3)-β-D-glucan, an essential cell wall polysaccharide that provides structural integrity, hence leading to the loss of cell wall integrity and severe cell wall stress. These drugs are very safe, possibly because they have a fungal-specific target, with no analogues in animal cells. Despite their potent activity against *Candida* spp. and *Aspergillus* spp., these compounds are completely ineffective against *Cryptococcus* spp. and *Trichosporon*. In this context, echinocandins have emerged as an important treatment option for candidiasis, especially because of the increasing prevalence of azole resistance (Table 2.1).

2.2 Antifungal Susceptibility Testing

Antimicrobial susceptibility testing (AST) of microrganisms has been performed primarily to predict the outcome of antimicrobial therapy in individual patients and to collect local data on antimicrobial susceptibility in order to support the use of empirical therapy. Moreover, AST has also been performed to detect organisms that carry especially unwanted resistance mechanisms, so that infection control and/or public health measures can be taken. Finally, considering that the world is headed toward a "post-antibiotic era," the interest in performing AST has increased, aiming at measuring the dynamics of the development of resistance and the potential effect of countermeasures to slow or reverse the emergence of resistant strains.

However, one of the greatest obstacles of developing methodologies to evaluate the antimicrobial susceptibility of microrganisms is to establish standardized protocols to obtain reproducible results that can be compared between different institutions. This is of utmost importance, as susceptibility results should be comparable all across the world, especially for epidemiological and surveillance purposes.

The need to develop reliable assays to evaluate the susceptibility of fungal isolates to antifungal drugs arose with the epidemics of AIDS, in the beginning of the 1980s, when it was observed that several AIDS patients with oro-esophageal candidiasis did not respond to antifungal therapy. In 1985, the Mycology committee of the Clinical Laboratory Standards Institute (CLSI), formerly known as the National Committee for Clinical Laboratory Standards (NCCLS), observed that 20% of the member laboratories started to perform antifungal susceptibility assays, especially with *Candida* spp., but they were obtaining discrepant results. Thus, the CLSI decided to develop and standardize a reproducible and feasible methodology to evaluate the antifungal susceptibility of clinical isolates for routine laboratories in the United States.

The same problem was observed in Europe, where national agencies from six countries (France, United Kingdom, Netherlands, Germany, Sweden, and Norway) would recommend different methodologies, leading to low reproducibility and comparability. Hence, in 1997, the European Society of Clinical Microbiology and Infectious Diseases (ESCMID) created the European Committee on Antimicrobial Susceptibility Testing (EUCAST) to form a committee that followed similar lines to those of the CLSI, by involving representatives of the national agencies that were already acting on the development of AST to standardize the methodologies used across Europe.

These initiatives (CLSI and EUCAST) led to the development of standardized protocols to reduce inter-institutional variability and increase comparability across different countries. Both organizations, CLSI and EUCAST, have elaborated protocols describing the different AST methodologies they recommend. CLSI's and EUCAST's latest documents were released in 2017, including one document for AST with yeasts, more specifically *Candida* spp. and *Cryptococcus* spp., and another for AST with filamentous fungi (conidium-forming molds).

2.2.1 Methods for antifungal susceptibility testing

Some methods have been described for evaluating the antimicrobial susceptibility of fungi, which can be divided into qualitative methods, when the concentration of drug required to inhibit fungal growth is not determined, or quantitative methods, when the minimum inhibitory concentration (MIC) of the tested drug can be determined.

The first described methodology for antifungal susceptibility testing was the disk-diffusion assay, which has been mainly used for yeasts similar to the one used for the evaluation of the antimicrobial susceptibility of bacteria, whose basic principle is the measurement of a zone of growth inhibition around a disk impregnated with a pre-established amount of drug. The greater the inhibition zone, the more susceptible the isolate to the tested drug. The values are expressed as a diameter (mm) of the inhibition zone and do not reflect the concentration of drug needed to inhibit microbial growth, therefore yielding qualitative results. This methodology has been mainly used for *Candida* spp. and some filamentous fungi, like *Aspergillus* spp. Moreover, epsilometer assays have been developed, which use strips impregnated with a decreasing drug gradient instead of using antifungal disks, allowing the estimation of antifungal MIC values. However, these values are an estimate; thus, the epsilometer test is considered a semi-quantitative method. Figure 2.1 illustrates the main methods used for antifungal susceptibility testing.

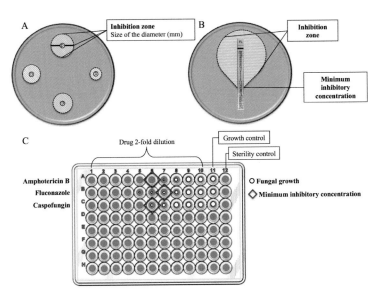

Figure 2.1   Main methods used for the performance of antifungal susceptibility assays. (A) Disk diffusion: the antifungal drug from the antifungal disk (D) diffuses within the agar, forming an inhibition zone, whose diameter is measured, expressed in millimeters, and its size allows the classification of the isolates into susceptible, intermediate, or resistant. This is a qualitative method, as the drug concentration cannot be determined. (B) Epsilometer method: the antifungal strip is impregnated with decreasing drug concentrations and drug diffusion within the agar leads to the formation of a drop-shaped inhibition zone that encounters the antifungal strip. The concentration at which the inhibition zone encounters the strip represents the estimated minimum inhibitory concentration (MIC) of the drug; hence, it is a semi-quantitative method. (C) Broth microdilution method: antifungal drugs are added at decreasing drug concentrations in order to find the lowest concentration capable of inhibiting fungal growth, defined as the MIC. It is the gold-standard quantitative method for antifungal susceptibility testing. Growth control: contains nutrient broth and fungal inoculum; Sterility control: contains only nutrient broth and/or antifungal drugs; Drug 2-fold dilution: dilution is performed from well #1 (highest concentration) to well #10 (lowest concentration). Amphotericin B: tested at a concentration range of 0.03125–16 μg/mL; the MIC (100% of growth inhibition) is 0.5 μg/mL (well #6). Fluconazole: tested at 0.125–64 μg/mL; the MIC (50% of growth inhibition) is 1 μg/mL (well #7). Caspofungin: tested at 0.03125–16 μg/mL; the MIC (50% growth inhibition) is 0.5 μg/mL (well #6).

The real quantitative methods are those that test serial drug dilutions in order to find the concentration that inhibits fungal growth. These methods are generically known as agar or broth dilution assays. Agar dilution assays can be carried out in Petri dishes containing nutrient agar and increasing drug concentrations. Broth dilution assays, on the other hand, are performed in glass slants, using large volumes of growth medium and drug, or in microtiter plates (broth microdilution), miniaturizing the procedures, making them cheaper and more practical (Figure 2.1).

Among the available methodologies used for the evaluation of the antifungal susceptibility, the broth microdilution is the gold standard. This method consists of exposing a standardized fungal inoculum to increasing 2-fold concentrations (10 different concentrations) of the antifungal drugs, using a 96-well U-bottomed or flat-bottomed polystyrene microtiter plate (Figure 2.1). Both institutions mainly recommend the use of RPMI 1640 broth, with or without glucose supplementation, as growth medium. Moreover, the inclusion of some control check-points is essential to ensure test reliability: (1) drug-free growth control well for each tested isolate, to warrant fungal viability; (2) fungus-free sterility control wells, to warrant the sterility of the reagents and procedures; and (3) quality control fungal strains, to validate the effectiveness of the tested antifungal drugs. Table 2.2 shows a parallel between the methodologies recommended by CLSI and EUCAST.

At the end, these recommended methodologies allow the determination of the minimum inhibitory concentration for the tested antifungals, based on the reading breakpoints established by each organization. MICs can be visually or spectrophotometrically read and are, basically, defined as the lowest drug concentration capable of inhibiting 100% of fungal growth, for drugs with fungicidal activity, such as amphotericin B; or the lowest concentration capable of partially (50%–80%) inhibiting fungal growth, for drugs with fungistatic activity, as azoles and echinocandins. It is important to emphasize that there are some methodological differences between the protocols established by CLSI and EUCAST, especially involving inoculum concentration, incubation period, and broth composition; however, great efforts have been made to harmonize the obtained results and interpretation (Table 2.2).

2.2.2 Minimum inhibitory concentration interpretation: Clinical breakpoints vs. epidemiological cutoff values

After obtaining the antifungal MIC values against the tested isolates, they can be classified as susceptible, intermediate, or resistant, according to the interpretive or clinical breakpoints established by CLSI and EUCAST. More recently, the definitions of these categories were revisited and are now described as follows: (1) *susceptible (S)*, when the level of antimicrobial activity (determined by the antifungal MIC) is associated with a high likelihood of therapeutic success; (2) *intermediate (I)*, when the level of antimicrobial activity (determined by the antifungal MIC) is associated with a high likelihood of therapeutic success, but only if a higher dosage (higher doses, higher frequency) of the agent than normal can be used or if the antifungal agent is physiologically concentrated at the site of infection; and (3) *resistant (R)*, when the level of antimicrobial activity (determined by the antifungal MIC) is associated with a high likelihood of therapeutic failure. Based on these definitions, it becomes clear that the establishment of clinical breakpoints depends on a trinomial interaction that involves (1) fungal species, (2) the used/ tested antifungal drug, and (3) the clinical outcome of the patients.

Hence, the main objective of performing AST is to predict the response of a given fungal isolate to antifungal therapy. However, similarly to what has been described for susceptibility testing of bacteria, the 90–60 rule also applies for AST with fungi. This rule establishes that 90% of the infections caused by susceptible microorganisms *in vitro* and 60% of the infections caused by resistant microorganisms *in vitro* respond well to *in vivo* antimicrobial therapy. Therefore, when clinical breakpoints are consistently established, even though there is no direct correlation between *in vitro* and *in vivo* results, the performance of AST can guide the institution of the antifungal therapy, leading the patient to a 90% chance of therapeutic success.

Nonetheless, determining reliable clinical breakpoints is still one of the greatest bottlenecks of antifungal susceptibility testing, as their establishment requires knowledge of the MIC distribution for each antifungal agent and fungal species, pharmacodynamic and phamacokinetic parameters of the drug, and clinical outcome, taking into account the relation between these data and the antifungal MIC values. In addition, several factors influence the clinical response, such as the

Table 2.2   Methodological Parallel between the Protocols Proposed by the Clinical Laboratory Standards Institute (CLSI) and the European Committee on Antimicrobial Susceptibility Testing (EUCAST) for the Performance of Broth Microdilution Assays

Steps	CLSI	EUCAST
Polystyrene plate	96-well U-bottomed	96-well flat-bottomed
Broth	RPMI 1640 (1×)	RPMI 1640 + 2% glucose (2×)
Fungal inoculum	*Yeasts*: 0.5–2.5 × 10³ cfu/mL	*Yeasts*: 0.5–2.5 × 10⁵ cfu/mL
	Molds:	*Molds*: 1–2.5 × 10⁵ cfu/mL
	Non-dermatophytes: 0.4–5 × 10⁴ cfu/mL	
	Dermatophytes: 1–3 × 10³ cfu/mL	
Incubation	*Yeasts*:	*Yeasts*:
	Candida spp.: 35°C, 24 h	*Candida* spp.: 35°C–37°C, 24 h
	Cryptococcus spp.: 35°C, 48–72 h	*Cryptococcus* spp.: 35°C, 48 h (30°C, if no growth at 35°C)
	Molds: 35°C, 24–72 h	*Molds*: 35°C, 24–72 h
Reading	*Yeasts and molds*: visual	*Yeasts*: spectrophotometric
		Molds: visual
Reading breakpoint (MIC or MEC determination)	*Yeasts*:	*Yeasts*:
	Amphotericin B: 100% inhibition	Amphotericin B: ≥90%[a] inhibition in absorbance
	Azoles, echinocandins, flucytosine: 50%[a] inhibition	Azoles, echinocandins, and flucytosine: 50%[a] inhibition in absorbance
	Molds:	*Molds*:
	Non-dermatophytes:	Amphotericin B and azoles[b]: 100% inhibition
	Amphotericin B and azoles: 100% inhibition	Echinocandins: abnormal growth (MEC)
	Fluconazole and flucytosine: 50% inhibition	
	Echinocandins: abnormal growth (MEC)	
	Dermatophytes:	
	Azoles, terbinafine, griseofulvin, and ciclopirox: 80%[a] inhibition	
Quality control strains	*Candida parapsilosis* ATCC 22019	
	Candida krusei ATCC 6258	
Sterility control	Yes	
Growth control	Yes	

Abbreviations: MIC, minimum inhibitory concentration, lowest drug concentration capable of inhibiting fungal growth; MEC, minimum effective concentration, lowest drug concentration capable of causing abnormal growth (short, branched hyphae), used only for echinocandins against filamentous fungi.

[a]   Partial (50%, 80%, or 90%) inhibition is always in relation to the turbidity or absorbance of the drug-free growth control.

[b]   EUCAST does not recommend the use of fluconazole or flucytosine against molds.

host immune status and site of infection. Thus, the establishment of clinical breakpoints depends on the observation and follow-up of several cases of infections caused by each fungal species, which makes it expensive and challenging, especially for fungal infections with low prevalence or those that mainly affect immunosuppressed individuals, as their debilitated immune status will most likely jeopardize clinical outcome.

Based on what was explained above, it is understandable why clinical breakpoints have only been established for a few drugs (azoles and echinocandins) against some *Candida* and *Aspergillus* species. These obstacles led to the questioning of the applicability of performing AST with fungal isolates, as MICs could not be correlated or further interpreted. Apparently, no biological meaning could be given to antifungal MICs. Because of that, the institutions started using a binomial approach for further interpretation of the obtained MIC data, one that only depends on (1) the population of a given fungal species, and (2) the tested antifungal drug. This approach allows an epidemiological analysis of the antifungal MICs against this fungal population, based on premises of population statistics, since antifungal MIC distribution against a given species tends to follow the Gaussian distribution. Figure 2.2 shows examples of antifungal MIC distribution and ECV determination.

This frequency distribution analysis allowed the establishment of the epidemiological cut-off value (ECV or ECOFF) for the tested antifungal drug, which classifies the isolates of a given fungal species into wild-type, when the isolates are susceptible to antifungal concentrations lower than or equal to the ECV (MIC ≤ ECV), or non-wild type, when the isolates are susceptible to antifungal concentrations higher than the ECV (MIC > ECV). Therefore, considering that MIC

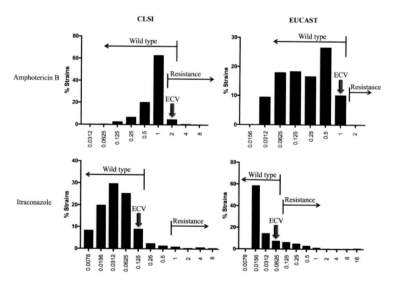

Figure 2.2   Frequency distribution of the minimum inhibitory concentrations (MICs—μg/mL) for amphotericin B and itraconazole against *Candida albicans*, obtained by the broth microdilution method, according to the protocols described by the Clinical Laboratory Standards Institute (CLSI) and the European Committee on Antimicrobial Susceptibility Testing (EUCAST). ECV: epidemiological cutoff value, antifungal MIC that divides the tested microbial population into wild-type (MIC ≤ ECV) and non-wild type (MIC > ECV). Clinical breakpoints used for the construction of these graphs were as follows: Amphotericin B: MIC >1 μg/mL indicates resistance, according to both CLSI and EUCAST; Itraconazole: MICs >0.5 μg/mL (CLSI) or >0.0625 μg/mL (EUCAST) indicate resistance. Notice that wild-type strains are mostly classified as susceptible strains, while non-wild type strains may be classified as susceptible or resistant, depending on the analyzed drug. Graphs created with data obtained from Pfaller et al. (2010, 2012) and EUCAST ECOFF database (https://mic.eucast.org/Eucast2/).

distribution follows the Gaussian distribution, the typical wild-type MIC distribution covers three to five twofold dilutions surrounding the modal MIC. The upper limit of the wild-type population is the ECV, which generally encompasses at least 95% of the isolates within the specific fungal population (Pfaller et al., 2010) (Figure 2.2).

But what is the biological meaning of categorizing an isolate as wild-type or non-wild type for a given antifungal drug? When the isolate is wild-type, it means that its susceptibility features to the antifungal drug are within the expected behavior for 95% of the isolates belonging to the same fungal species. On the other hand, when the isolate is classified as non-wild type, it means that its susceptibility features are not within the expected for the general population, so it may have gone through genetic/phenotypical changes. Therefore, a *wild-type (WT)* isolate is characterized by the absence of phenotypically detectable acquired and mutational resistance mechanisms to the antifungal agent in question, while a *non-wild type (NWT)* isolate is characterized by the presence of phenotypically detectable acquired or mutational resistance mechanisms to the agent in question.

Even though ECVs do not directly predict the response to antifungal therapy, they have proven to be useful to monitor the expression of resistance mechanisms and hence the emergence of antifungal resistance. Moreover, ECVs also allow the surveillance of the susceptibility of a given fungal species to a specific antifungal drug throughout time, enabling a better understanding of the epidemiological behavior of fungi, especially in the presence of environmental selective pressures.

Technically, ECVs may be established for any antifungal drug against any fungal species, as long as there are MIC data from different institutions, obtained through the same standardized methodology, against a minimum of 100 isolates. It is important to emphasize that a greater amount of MIC data and participating institutions lead to more reliable ECVs. Both CLSI and EUCAST have produced documents with the recommendations for the establishment and interpretation of antifungal ECVs. In this context, ECVs for several fungal species, including those of the genus *Candida*, *Cryptococcus*, *Malassezia*, *Aspergillus*, *Fusarium*, *Sporothrix*, and *Histoplasma*, have been established.

2.2.3 Automation in antifungal susceptibility testing

The same systems used for the phenotypical identification of fungal species, such as the Vitek 2 System (BioMérieux, France), can also be used to evaluate the antifungal susceptibility of the isolates. These devices are only able to identify and analyze yeast isolates belonging to the species they contain in their database. Thus, they can reliably perform antifungal susceptibility assay with *Candida albicans*, *Candida tropicalis*, *Candida parapsilosis*, *Candida glabrata*, and *Candida krusei*, and few others. The methodology applied by these devices is a broth microdilution using only three to five drug concentrations for each tested drug, which allows the evaluation of fungal growth inhibition and estimation of MIC values. These analyses are performed in manufactured cards or panels, which contain the drugs to be tested, that is, amphotericin B, flucytosine, fluconazole, voriconazole, caspofungin, and micafungin. The results are then released as MIC values, followed by the susceptibility category (S, I, or R) and epidemiological category (WT or NWT). These automated methods are very practical for the most prevalent yeast pathogens, but their major drawback is the lack of flexibility to perform the assays, as the tested species must be in the system database and the products and reagents used are developed, industrialized, and supplied by the companies.

Bibliography

CLSI. *Principles and Procedures for the Development of Epidemiological Cutoff Values for Antifungal Susceptibility Testing*. 1st ed. CLSI guideline M57. Wayne, PA: Clinical and Laboratory Standards Institute; 2016.
CLSI. *Reference Method for Broth Dilution Antifungal Susceptibility Testing of Yeasts*. 4th ed. CLSI standard M27. Wayne, PA: Clinical and Laboratory Standards Institute; 2017.
CLSI. *Reference Method for Broth Dilution Antifungal Susceptibility Testing of Filamentous Fungi*. 3rd ed. CLSI standard M38. Wayne, PA: Clinical and Laboratory Standards Institute; 2017.

CLSI. *Epidemiological Cutoff Values for Antifungal Susceptibility Testing*. 2nd ed. CLSI supplement M59. Wayne, PA: Clinical and Laboratory Standards Institute; 2018.

Espinel-Ingroff, A., Turnidge, J. 2016. The role of epidemiological cutoff values (ECVs/ECOFFs) in antifungal susceptibility testing and interpretation for uncommon yeasts and moulds. *Rev Iberoamer Micol* 33: 63–75.

EUCAST. Data from the EUCAST MIC distribution website, last accessed 22 April 2019. http://www.eucast.org

EUCAST. Breakpoint tables for interpretation of MICs. Version 9.0, valid from 2018-02-12. http://www.eucast.org/fileadmin/src/media/PDFs/EUCAST_files/AFST/Clinical_breakpoints/Antifungal_breakpoints_v_9.0_180212.pdf.

EUCAST DEFINITIVE DOCUMENT E.DEF 7.3.1. Method for the determination of broth dilution minimum inhibitory concentrations of antifungal agents for yeasts, 2017. http://www.eucast.org/fileadmin/src/media/PDFs/EUCAST_files/AFST/Files/EUCAST_E_Def_7_3_1_Yeast_testing__definitive.pdf.

EUCAST DEFINITIVE DOCUMENT E.DEF 9.3.1. Method for the determination of broth dilution minimum inhibitory concentrations of antifungal agents for conidia forming moulds, 2017. http://www.eucast.org/fileadmin/src/media/PDFs/EUCAST_files/AFST/Files/EUCAST_E_Def_9_3_1_Mould_testing__definitive.pdf.

Geddes-McAlister, J., Shapiro, R. S. 2019. New pathogens, new tricks: Emerging, drug-resistant fungal pathogens and future prospects for antifungal therapeutics. *Ann NY Acad Sci* 1435: 57–78.

Gonzalez, J. M., Rodriguez, C. A., Agudelo, M., Zuluaga, A. F., Vesga, O. 2017. Antifungal pharmacodynamics: Latin America's perspective. *Braz J Infect Dis* 21: 79–87.

Kahlmeter, G. 2015. The 2014 Garrod Lecture: EUCAST—Are we heading towards international agreement? *J Antimicrob Chemother* 70: 2427–2439.

Lewis, R. E. 2011. Current concepts in antifungal pharmacology. *Mayo Clin Proc* 86: 805–817.

Lockhart, S. R., Ghannoum, M. A., Alexander, B. D. 2017. Establishment and use of epidemiological cutoff values for molds and yeasts by use of the Clinical and Laboratory Standards Institute M57 Standard. *J Clin Microbiol* 55: 1262–1268.

Pfaller, M. A., Boyken, L., Hollis, R. J. et al. 2010. Wild-type MIC distributions and epidemiological cutoff values for the echinocandins and *Candida* spp. *J Clin Microbiol* 48: 52–56.

Pfaller, M. A., Espinel-Ingroff, A., Canton, E. et al. 2012. Wild-type MIC distributions and epidemiological cutoff values for amphotericin b, flucytosine, and itraconazole and *Candida* spp. as determined by CLSI broth microdilution. *J Clin Microbiol* 50: 2040–2046.

Robbins, N., Wright, G. D., Cowen, L. E. 2016. Antifungal drugs: The current armamentarium and development of new agents. *Microbiol Spectrum* 4: FUNK-0002-2016.

Tabara, K., Szewczyk, A. E., Bienias, W. et al. 2015. Amorolfine vs. ciclopirox—Lacquers for the treatment of onychomycosis. *Postepy Dermatol Alergol* 32: 40–45.

3
Candida spp.

Silviane Praciano Bandeira, Glaucia Morgana de Melo
Guedes, and Débora de Souza Colares Maia Castelo-Branco

Contents

3.1 General aspects

Yeasts of the genus *Candida* are part of the microbiota of humans and animals, colonizing anatomical sites such as the oral cavity, vaginal mucosa, and intestinal tract. It is estimated that approximately 75% of the world's population is colonized in the oral cavity with some *Candida* species. In most individuals, these yeasts remain commensal throughout life, with no risk to health. In certain situations, however, they can cause infections with a wide spectrum of presentations, from superficial cutaneous conditions to severe and disseminated conditions that can lead to death (Vale-Silva et al., 2012). This genus is also present commensal of mucosal surfaces of several animal species, including dogs, cats, goats, sheep, and birds. In addition, these species are also potentially pathogenic for animals, depending on several predisposing conditions (Castelo-Branco et al., 2013).

The species of this genus are the main fungal pathogens, making up the fourth leading cause of nosocomial bloodstream infections (Vale-Silva et al., 2012). In some hospitals, this situation is even more worrying, since these yeasts are among the three main pathogens of healthcare-related infections in several anatomical sites.

The occurrence of fungal infections has increased in recent decades. This increase is due to the emergence of a large population of individuals susceptible to that type of infections, as a consequence of the emergence of the human immunodeficiency virus (HIV) in the 1980s, among other factors. The development of broad and effective antiretroviral therapy has ensured longer survival for HIV-infected patients. Similarly, the improvement of diagnostic techniques for neoplastic conditions and the development of a potent therapeutic arsenal for these patients guarantee greater survival. These situations increase the likelihood of developing fungal infections, increasing the number of cases of different types of mycoses (Ko et al., 2015).

Factors associated with the emergence of infections in humans and other animals can be related to host or pathogenic potential of *Candida* strains. In human cases, patient-related factors include suppressed cell immunity, decompensated diabetes mellitus, extensive and prolonged antibiotic therapy, age-associated debilitation, use of invasive devices, intestinal mucosal damage, prolonged hospitalization, use of corticosteroids, malignant neoplasms, transplants, and autoimmune diseases (Giacomazzi et al., 2016). Regarding issues related to the microorganism, important factors for pathogenicity are the activity of molecules associated with adhesion and invasion, presence of hydrolytic enzymes that promote invasiveness, ability to shift between blastoconidia and filamentous structures, and ability to form biofilms, among others. It should be mentioned that the mere occurrence of these factors does not guarantee the infectious process. The establishment of *Candida* sp. depends on an imbalance in the parasite-host relationship, providing a microenvironment for the invasion and reproduction of the microorganism. In summary, the infectious process is opportunistic (Ko et al., 2015).

Of the species of *Candida* genus, slightly more than a dozen are known to be pathogenic to humans and animals. The main ones are *Candida albicans, Candida tropicalis* members of the *Candida parapsilosis* complex, *Candida glabrata, Candida guilliermondii, Candida krusei*, and *Candida dubliniensis*. *C. albicans* is recognized as the most pathogenic species and is the main etiological agent of superficial and deep fungal infections (Pfaller & Diekema, 2010). Despite this, non-*Candida albicans* species have become important colonizers and pathogens and are often isolated from biological samples from humans and animals. Although the most diagnosed species remains the same, studies on this fungal genus are very dynamic and new information has been continually emerging (Chowdhary et al., 2017).

In 2016, the Centers for Disease Control and Prevention (CDC) published a new species emergence alert, warning against the impressive resistance phenotype of this species. *C. auris* was first described in 2009 in Japan, where it was isolated from a patient's external ear canal (Satoh et al., 2009). In Brazil, in 2017, the National Health Surveillance Agency (ANVISA) published a Risk Communication about this species and emphasized that identification based on conventional techniques may be misleading (Kathuria et al., 2015), leading to the identification of less common species such as *C. haemulonii, C. famata, C. sake, C. catenulata, C. lusitaniae, C. guilliermondii, Rhodotorula glutinis*, and *Saccharomyces cerevisiae*, depending on the identification platform used.

Some other species of the genus have undergone reclassification based on molecular biology techniques, and were grouped as cryptic species (Sullivan et al., 1995). These species can only be reliably identified when morphology-based techniques are not used alone, requiring the use of molecular techniques (Bickford et al., 2007). The yeasts of the *Candida parapsilosis* complex are classic examples of cryptic species (Tavanti et al., 2005), as well as *C. albicans* and *C. dubliniensis* (Fricke et al., 2010).

The fast and correct laboratory identification of *Candida* species can play an important role in the control of infections, decreasing mortality rates (Neppelenbroek et al., 2014), since some strains are intrinsically resistant to antifungals, such as *C. glabrata* and *C. krusei*, which are commonly resistant to fluconazole, and *C. auris*, a multidrug-resistant species (Chowdhary et al., 2017).

3.2 Laboratory diagnosis

Sample collection for laboratory diagnosis of *Candida* spp. depends on the infection site. In cutaneous mucosal lesions, most clinical specimens are collected using sterile swabs that must be immediately processed; otherwise, they must be packed in a sterile transport medium or saline, since *Candida* species are susceptible to desiccation. In dry lesions, as in onychomycosis, nail scrapings obtained from the region of progression, where there is a confluence of healthy with diseased tissue, may be used. In systemic candidiases, clinical specimens such as blood, bronchoalveolar aspirate, bronchial lavage material, urine, and biopsy of several organs can be used for diagnosis. The guidelines regarding the collection and transportation of different clinical samples were described in Chapter 1.

The classic laboratory diagnosis of microorganisms involves techniques based on phenotypic characteristics. Identification can also be based on observation of the microorganisms in a clinical material, for example. The microbiological routine is usually initiated by microscopic analysis. In addition to allowing qualitative evaluation of the clinical sample, microscopy allows visualization of characteristic structures of a given group of microorganisms, including bacteria or fungi.

3.2.1 Microscopy

For mycological diagnosis, the microscopic observation of pathogens can be performed by preparing slides with clarifying agents, such as sodium or potassium hydroxide, at concentrations of 10%–40%, in the case of materials that need clarification, like skin, nail scrapings, and scalp samples. Direct examination is also possible, without clarifying agent. Several staining techniques can be used in mycology, such as impregnation with silver. In this case, the fungal structures acquire a brownish coloration on a green background. All these techniques involve optical microscopy.

In addition, fluorescence microscopy can also be used, by using calcofluor dye, which highlights the fungal structures with a bluish-white glow by binding to the chitin of the fungal wall. Occasionally, Gram staining can be used for some clinical samples, when yeasts of the genus *Candida* appear as Gram-positive round or oval structures, even though budding blastoconidia or tubular structures, such as hyphae or pseudohyphae, may also be seen.

3.2.2 Culture

Growth in culture media is the gold standard for identifying any microorganism, and obtaining pure cultures from viable clinical samples is the major goal of microbiology laboratories.

The culture media classically used in mycology are Sabouraud agar, Sabouraud agar with chloramphenicol, and Sabouraud with chloramphenicol and cycloheximide (also known as Mycosel agar). Sabouraud agar is the classic medium, which is supplemented with peptone and dextrose, allowing the growth of all cultivable fungi. Chloramphenicol is an antibiotic used to prevent the growth of competing bacteria that may be in the clinical sample, and cycloheximide inhibits the growth of contaminating fungi. Most yeast species of the genus *Candida* grow on all these culture media, presenting creamy texture, yellowish-white colonies, regular borders, and uniform relief after incubation at room temperature or 35°C for 3 to 5 days.

3.3 Identification methods

3.3.1 Conventional

The macroscopic aspect of the colonies is complemented by micromorphological and physiological characteristics to consolidate identification. Table 3.1 summarizes different phenotypic methods that may be valuable for the identification of different *Candida* species when used together.

In the laboratory routine, chromogenic media are used, on which several species of *Candida* present species-specific colors that may vary according to the brand of the chromogenic medium. ChromAgar *Candida* (Figure 3.1A) is the most widely used chromogenic medium in mycology. On this agar, the colonies of *C. albicans* and *C. dubliniensis* turn green, *C. tropicalis* assumes shades of blue, and *C. parapsilosis* species complex are pale pink to lavender. *C. krusei* also presents a pinkish shade with a rough colony. This identification is only presumptive, requiring the use of other methods to confirm the species (Kurtzman et al., 2011).

In the *Candida* genus, the ability to generate true hyphae from blastoconidia is a peculiar feature of some species, such as *C. albicans, C. dubliniensis*, and *C. africana*. This feature can be verified by the germ tube test. A small aliquot of the colony to be tested is incubated with 0.5 mL of human or horse serum at 37°C for up to 3 hours. At 30-minute intervals, an aliquot of the fungal suspension is submitted to microscopic observation under magnifications of 100× and 400×. When positive, germ tube formation is seen, which is characterized by the presence of a parallel-walled tubular structure protruding from a yeast cell.

Dalmau slide culture enables the micromorphological study of *Candida* spp. by growing the isolates in Petri dishes containing cornmeal agar or rice agar supplemented with Tween 80 (Figure 3.1B and C). For this purpose, the yeast colony is drawn through the agar, making three parallel lines, on which a sterile glass coverslip is placed, after flamed and allowed to cool. Under the coverslip, a microenvironment is generated with reduced oxygen concentration, which leads to the formation of typical structural arrangements, including resistance structures, like chlamydoconidia, typical of *C. albicans* and *C. dubliniensis*. After an incubation at 28°C for 3–5 days, the plates are observed under optical microscopy at magnifications of 100× and 400×.

Table 3.1 Conventional Tests to Identify Major *Candida* spp. of Medical Importance

Yeast Species	CHROM	Fermentation										Assimilation															Microculture
		Glu	Suc	Gal	Tre	Mal	Raf	Lac	Inu	Rha	Ara	Glu	Suc	Lac	Gal	Raf	Ino	Xyl	Cel	Tre	Dul	Mal	Mel	N	U	GT	
Candida albicans	Gr	+	>	>	>	+	−	−	−	−	>	+	>	−	+	−	−	+	−	>	−	+	−	−	−	+	Presence of clusters of blastoconidia and chlamydoconidia
C. glabrata	W, Pr, Pi	+	−	−	>	−	−	−	−	−	−	+	−	>	−	−	−	−	−	>	−	−	−	−	−	−	Only small blastoconidia
C. guilliermondii	Pi, Lil	+	+	>	+	−	+	−	+	>	+	+	+	−	+	+	−	+	+	+	>	+	+	−	−	−	Blastoconidia in constrictions of very thin pseudomycelium
C. krusei	Pi	+	−	−	−	−	−	−	−	−	−	+	−	−	−	−	−	−	−	−	−	−	−	−	−	−	Elongated blastoconidia grouped in the constrictions of pseudomycelium
C. parapsilosis	W, Lil	+	>	>	>	>	−	−	−	−	+	+	+	−	+	−	−	+	−	+	−	+	−	−	−	−	Presence of giant cells and curved and thin pseudohyphae
C. tropicalis	B	+	>	+	>	+	−	−	−	−	−	+	>	−	+	−	−	+	>	+	−	+	−	−	−	−	Blastoconidia in single or branched chains

Abbreviations: v = variable; + = positive; − = negative; N = nitrate; U = urease; GT = germ tube; Gr = green; W = White; Pr = Purple; Pi = Pink; B = blue; Lil = lilac; Glu = glucose; Suc = sucrose; Gal = galactose; Tre = trehalose; Mal = maltose; Raf = raffinose; Lac = lactose; Rha = rhamnose; Ara = arabinose; Ino = inositol; Xyl = xylose; Cel = cellobiose; Dul = dulcitol; Mel = melibios.

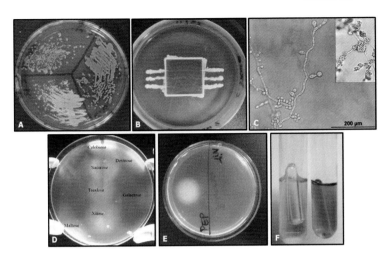

Figure 3.1 Physiological and micromorphological tests to identify *Candida* spp. (A) ChromAgar *Candida* presenting different species; (B) Dalmau slide culture; (C) Microscopy of the slide culture of *C. albicans*; (D) Carbohydrate assimilation showing fungal growth and the formation of halos under different carbon sources, indicating the assimilation of the carbohydrates; (E) Nitrogen assimilation showing fungal growth and halo formation due to the assimilation of peptone as nitrogen source; (F) Carbohydrate fermentation where production of carbon dioxide is observed, change of pH of the medium, and change of coloration from green to yellow in positive samples.

The physiological properties of yeast isolates are also important for species identification and are often necessary to reliably identify microorganisms. In this context, the ability of a given isolate to assimilate carbohydrates and nitrogen provides important information for species identification. Carbohydrate assimilation is the ability to use different sugars, such as maltose, sucrose, galactose, lactose, trehalose, melibiose, arabinose, cellobiose, xylose, raffinose, dulcitol, rhamnose, inulin, mannitol, and inositol, among others, as sources of carbon under aerobic conditions. In this assay, dextrose is used as positive control for sugar assimilation (Figure 3.1D). Similarly, the nitrogen assimilation test evaluates the ability of an isolate to use inorganic nitrogen, commonly potassium nitrate, as a nitrogen source, and peptone, an organic compound, is used as positive control for nitrogen assimilation (Figure 3.1E). These assays are performed on carbon-free or nitrogen-free agars for carbon or nitrogen assimilation, respectively. Plates are incubated at 25°C–30°C for 24–96 hours. Positivity for assimilation assays is shown by fungal growth under the tested sugar or nitrogen source, presenting the formation of a halo under the tested compounds.

The ability to ferment different carbohydrates is also evaluated to identify *Candida* species. Fermentation tests are performed by inoculating a yeast suspension into test tubes containing culture medium with different sugars such as maltose, sucrose, galactose, lactose, and trehalose, and an inverted Durham tube is added to the test tubes (Figure 3.1F). Dextrose is used as positive control for sugar fermentation. Fermentation is observed through the production of carbon dioxide and pH change after incubation at 25°C–30°C for up to 28 days (Neppelenbroek et al., 2014).

3.3.2 Semiautomatic or automatic platforms

Identification tests can be performed by conventional manual methods. However, miniaturization technology has allowed the development of semiautomatic or automatic platforms to identify the main yeasts of clinical importance.

API 20C systems (BioMérieux, Marcy l'Étoile, France) are examples of semiautomatic identification platforms (Melhem et al., 2014). The API 20C Aux system consists of a panel containing 20 wells with dehydrated substrates that allows the performance of 19 assimilation tests. After incubation at 30°C (24, 48, and 72 h), the turbidity of the wells can be evaluated and a numerical code is generated, which corresponds to a microorganism present in the database.

Among the main automatic identification systems used in laboratories are: VITEK (BioMérieux, Marcy l'Étoile, France), Phoenix (BD, New Jersey, EUA), and WalkAway (Baxter Diagnostic, Inc., West Sacramento, CA, EUA). The use of these tools in some clinical laboratories is still limited due to the high cost (Deak et al., 2015). For VITEK, cards containing 64 wells with 47 biochemical tests are used. The results are obtained in 18–24 hours and are classified based on the confidence level of identification (Deak et al., 2015). An additional benefit of the automation of these tests is the parallel performance of antifungal susceptibility assays with the definition of antifungal minimum inhibitory concentration (MIC), according to standards of international institutions.

3.3.3 Molecular identification

Molecular tests have gained importance because of the difficulty or inability of certain isolates to grow in artificial media or when conclusive phenotypic identification is not achieved. The application of molecular biology includes the use of panfungal primers, which may lead to the identification of the genus of fungal pathogens. From this, the amplified region is sequenced and compared to available databases, and the most probable identification is obtained. Another form of molecular differentiation is the sequencing of specific regions within each genus.

The application of polymerase chain reaction (PCR) techniques associated with restriction enzyme assays is useful for the differentiation of the cryptic species of the *C. parapsilosis* species complex (Tavanti et al., 2005), for example. Previously designated as *C. parapsilosis* groups II and III, the species *C. orthopsilosis* and *C. metapsilosis*, respectively, were described by Tavanti et al. (2005). According to the authors, several molecular approaches can be used to differentiate cryptic species. The amplification of the secondary alcohol dehydrogenase gene (SADH) with subsequent restriction with BanI enzyme has been an effective method and has been applied by different researchers in later studies (Brilhante et al., 2014).

Another example is the identification of the cryptic species *C. albicans* and *C. dubliniensis*. Since the description of *C. dubliniensis* by Sullivan et al. (1995), it has been known that these species share several phenotypic characteristics that may lead to misidentification. Although some phenotypic criteria may differ between species, such as growth at 42°C, macroscopic colony pattern on some culture media, and some biochemical profiles, the definitive differentiation of species usually occurs through molecular biology tools, such as PCR or real-time PCR (Ahmad et al., 2012; Fricke et al., 2010).

Besides DNA-based molecular methods, the study of proteomics by using mass spectrometry techniques is a tool currently used for the identification of microorganisms. It is also a molecular method, since it is based on the molecules expressed in the structure of the pathogen. The great advantage of this method is the time for identification, taking only a few minutes, while conventional phenotypic tests can take days or weeks depending on the metabolism of the studied microorganism.

MALDI-TOF technology is the leading example of this new tool. The acronym is derived from matrix-associated laser desorption ionization-time of flight. It is a system in which a small aliquot of a pure and isolated colony or clinical material such as blood culture is placed on a metal plate and covered by a polymeric matrix (Barberino et al., 2017). A laser beam radiates this sample and ionizes it, generating several molecules that migrate in an electrically charged vacuum tube. This migration occurs as a function of mass and charge of the molecules. Thus, lighter particles arrive faster at the end of the tube, where there is a detector. The expression "time of flight," which allows differentiation between the molecules, is derived from this. Based on the time of arrival of the particles, the equipment generates a graph of peaks that differ between species of fungi and bacteria. This graph is compared to the equipment's software database, leading to the identification of the microorganism. The whole process takes around 5–15 min (Kassim et al., 2017). The technology is already under use to identify species of yeast and filamentous fungi, relying on a large database that encompasses the main fungal pathogens (Rizzato et al., 2015).

Although this method has attracted many enthusiasts in the scientific world, it has two major drawbacks. First, the cost of the equipment is still impeditive for many institutions. However, after the machine is installed, the cost of the test per sample is considered lower than that of other automated methods (Rizzato et al., 2015). Second, and perhaps more important than cost, is the limitation to conduct antimicrobial susceptibility testing. There are no applications of this method to formulate antibiograms, although some studies provide evidence of resistance behavior by the proteomic molecular pattern. In other words, while a microorganism may be identified in a few minutes, the therapeutic options for the case will only be available through other techniques that require more time. In an epidemiologically known context, identification may already suggest therapeutic reactions. From the clinical point of view, this is perhaps the greatest application of the method.

3.4 Therapeutic options

The available therapeutic arsenal to treat fungal infections is very limited, especially in cases of systemic and severe mycoses. Another aggravating factor is that the suspicion of fungal infection is often late, so the treatment is usually delayed. With the possibility of resistance, the chances of therapeutic failure increase, which can compromise patient survival (Giacomazzi et al., 2016).

For fungal infections by the genus *Candida*, there are therapeutic options with different routes of administration, such as topical, oral, and parenteral use. Therapeutic choice depends on clinical presentation, severity, etiological definition of the disease, and patient-related characteristics. Superficial infections can be controlled by topical or oral medications. Deep-seated conditions demand a more aggressive therapeutic approach, requiring the use of parenteral (intravenous) drugs.

Azole derivatives are the most widely used group of antifungal drugs worldwide. Ketoconazole and triazole derivatives—fluconazole, itraconazole, voriconazole, posaconazole, and others—are the clinical representatives of this class of drugs. These drugs are synthetic compounds that act by inhibiting the biosynthesis of ergosterol, the main sterol of the fungal membrane. This inhibition is due to the enzymatic blockade of lanosterol-14-α-demethylase, which converts lanosterol to ergosterol. In *C. albicans* strains, this enzyme is encoded by the ERG11 gene (Akins, 2005). Therefore, by inhibiting this conversion process, these drugs cause structural fragility of the microorganism and accumulation of toxic precursors to the fungus, killing the pathogen (Morschhauser et al., 2007).

Polyene derivatives are also an important tool for the treatment of fungal infections. These drugs act by binding to the ergosterol of the lipid bilayer of the fungal cell membrane, disrupting the cell architecture. They lead to the formation of pores, through which ions and molecules important for the survival of the microorganism are lost, causing its death (Akins, 2005). This group includes nystatin and amphotericin B. The use of the former is limited to cutaneous or mucosal application. Amphotericin B, in turn, is often the last resource for serious fungal infections, at the expense of high nephrotoxicity to the host, leading to severe hypokalemia and renal failure. Lipid formulations, the best-tolerated drug options, still present toxicity and are costly (Vincent et al., 2013).

Flucytosine, or 5-fluoro-cytosine, is a pyrimidine analog drug that inhibits fungal replication and impairs DNA transcription. Because of the high resistance rates due to modifications of important enzymes for its metabolism, the use of flucytosine is ideally performed in synergy with amphotericin B, allowing the use of smaller doses of amphotericin B and reducing its potential toxicity (Akins, 2005).

The most recently described class of antifungal drugs is the echinocandins. These drugs act quite peculiarly by preventing the synthesis of glucans in the fungal cell wall by inhibiting the enzyme 1,3-β-glucan synthase. This unique mechanism significantly reduces the occurrence of side effects, since it targets a structure that does not exist in the host cells. This class includes the drugs caspofungin, micafungin, and anidulafungin. They are expensive, but are a good therapeutic options in cases of *Candida* fungemia (Akins, 2005; Pfaller et al., 2012).

4
Cryptococcus

Luciana Trilles, Márcia dos Santos Lazéra, and Bodo Wanke

Contents

4.1 Introduction

Inhalation of spores or desiccated yeasts cells of *Cryptococcus neoformans* or *C. gattii* from the environment causes natural pulmonary infection that is usually regressive and unspecific. However, some lung infections can progress and spread hematogenously to other sites or organs. Cryptococcosis is a worldwide emerging invasive fungal infection that is capable of causing meningitis in immunocompromised as well as in apparently immunocompetent hosts. Outbreaks in tropical and temperate regions recently also became part of the complex epidemiology of this polymorphic mycosis. The agents are melanin-forming encapsulated yeasts that survive and adapt under different environmental conditions. Natural habitats and sources of infection are mostly associated with tree decay, dust in domestic environments, and bird excreta (Acheson et al., 2018; Brito-Santos et al., 2015; Kwon-Chung et al., 2014).

Cryptococcosis by *C. neoformans* is one of the most common life-threatening fungal infections in immunocompromised patients, particularly among people living with HIV. The latest estimates suggest that HIV-associated cryptococcal meningitis accounts for 150,000–200,000 deaths per year, mostly in sub-Saharan Africa, where the associated lethality rate remains at around 70%. Globally, cryptococcal meningitis was estimated to be responsible for 15% of AIDS-related deaths (Rajasingham et al., 2017). However, this picture is different according to the geographic region and the human development index. Recent data demonstrated that the annual rate of Cryptococcosis in HIV patients decreased markedly in North America (O'Halloran et al., 2017), while in developing countries, several studies highlighted the substantial ongoing burden of HIV-associated cryptococcal disease (Maziarz & Perfect, 2016; Rajasingham et al., 2017).

Other important risk factors for cryptococcosis are rheumatologic diseases, systemic lupus erythematosus, rheumatoid arthritis, corticosteroid and/or immunosuppressive therapies, diabetes mellitus, solid organ transplantation, decompensated chronic liver disease, malignant and lymphoproliferative disorders, renal failure and/or peritoneal dialysis, sarcoidosis, hyper-IgM syndrome or hyper-IgE syndrome, and treatment with monoclonal antibodies (etanercept, infliximab, alemtuzumab) (Maziarz & Perfect, 2016). It must be taken into consideration that *C. neoformans* is an important cause of cryptococcosis in apparently immunocompetent hosts in some regions, as in China (Chen et al., 2018).

First described in Central Africa, cryptococcosis by *C. gattii* predominates in apparently immunocompetent hosts in tropical regions, and since 1999 in temperate areas. In British

Columbia, Canada, the largest known outbreak of cryptococcosis in humans and animals was identified, and further expanded along the Northwest Pacific coast, reaching the United States. Outbreaks were also described in Australia, Brazil, and Italy. Cryptococcosis by *C. gattii* is largely endemic in the Amazon (North) and the semi-arid (Northeast) of Brazil (Acheson et al., 2018; Alves et al., 2016; Santos et al., 2008; Trilles et al., 2008).

Both infections, by *C. gattii* and *C. neoformans*, may occur in the same geographical area, such as the North and Northeast of Brazil, where 60%–88% of cryptococcosis by *C. gattii* were reported to occur in immunocompetent individuals, including children and adolescents (Martins et al., 2011; Santos et al., 2008). Nowadays, in this same region, cryptococcosis predominates in immunocompromised individuals, mostly caused by *C. neoformans* (Soares and Coutinho, personal communication), following the HIV trends.

The agents of cryptococcosis—*C. neoformans* and *C. gattii*—are sibling species that share the same common ancestor. Initially, the two etiologic agents were classified as only one species, but were distinguished by their capsular antigenic diversity (*C. neoformans* serotypes A and D, and *C. gattii* serotypes B and C) and other characteristics. The discovery of two different sexual forms, one for *C. neoformans* and the other for *C. gattii*, led to the recognition of two distinct species, confirmed later by whole-genome sequence data. Moreover, the two species have hidden diversity as well as hybrids intra- and interspecies. Phylogenetic analysis based on multilocus sequence typing (MLST) and amplified fragment length polymorphism (AFLP) showed that both *C. neoformans* and *C. gattii* strains were composed of multiple genetically diverse monophyletic clades (7–9). A recent study based on 115 isolates proposed to divide *C. neoformans* into two species and *C. gattii* into five species (Kwon-Chung et al., 2017). Considering that the naming of every clade as a separate species led to dramatic change in the naming of these medically important yeasts, a group of clinicians and researchers revised this issue and recommend the use of "*Cryptococcus neoformans* species complex" and "*C. gattii* species complex" as a practical intermediate step, rather than creating more species. This strategy recognizes genetic diversity without creating confusion and nomenclatural instability (Kwon-Chung et al., 2017). Researchers and clinicians should work together to achieve a reasonable degree of nomenclatural stability during the decades when large changes are unavoidable.

In the past, serotypes were identified by tests based on agglutination of cells against specific capsular antibodies, but the test is no longer commercially available or used in epidemiological studies. Currently, the *C. neoformans* species complex includes major molecular types named VN types that correspond to serotypes or varieties used within the extensive previous related literature. For example: VNI and VNII types correspond to serotype A, also named *C. neoformans* var. *grubii*; VNIV type corresponds to serotype D, named *C. neoformans* var. *neoformans* by some authors. In the same way, *C. gattii* species complex includes four major VG molecular types and two serotypes (B and C). The molecular VN-VG types are identified by the *URA5*-RFLP method, but some authors, to identify major AFLP types, also use the AFLP method (Trilles et al., 2003).

In brief, *C. neoformans* is cosmopolitan, affecting mainly immunocompromised individuals, especially patients with AIDS (Kwon-Chung & Bennett, 1992). On the other hand, *C. gattii* predominantly causes a primary infection in immunocompetent individuals, previously associated with tropical and subtropical climates, but now gaining prominence as an important cause of human and veterinary disease in temperate regions of North America (Lockhart et al., 2013).

4.2 Clinical diagnosis

Cryptococcosis manifests as a respiratory, disseminated, or meningeal disease. The pattern of cryptococcal pulmonary disease varies, ranging from asymptomatic to acute respiratory distress syndrome. The symptomatology is not specific, and thoracic discomfort and image finding in an immunocompetent host are not rare. Pulmonary cryptococcal disease in an immunocompetent host may manifest as a peripheral nodule, subpleural and sometimes eccentric, usually single, well defined, non-calcified, without evident pleural effusion. Pulmonary cryptococcosis should be considered on a differential diagnosis for nodular or mass-like lesions in chest radiograph, as it often mimics lung cancer (Kim et al., 2012).

It is important to emphasize that the isolation of *C. neoformans* or *C. gattii* from the tracheobronchial tree may rarely correspond to the colonization of this mucosa. Endobronchial lesions by *Cryptococcus* sp. and incipient pulmonary lesions should be routinely investigated by bronchofibroscopy, chest tomography, or magnetic resonance. Investigation of the cryptococcal antigen in the serum is essential, and, if positive, it denotes a potential invasive cryptococcosis.

In immunocompromised patients, such as those with AIDS with low CD4 counts, pulmonary cryptococcosis may present as a localized or diffuse bilateral infiltrate with a reticulonodular pattern and, more rarely, hilar lymph node enlargement. Due to the unspecific pattern and the prevalence of pulmonary tuberculosis in this risk group, only laboratorial exams will support the diagnosis.

The central nervous system (CNS) infection is the most important and fatal manifestation of cryptococcosis, predominating as meningoencephalitis, associated or not with cerebral nodules. Non-AIDS patients claim intense occipital headache, which does not respond to analgesics. At the beginning, it can mimic sinusitis, but the persistent headache is remarkable. Visual complaints such as blurred vision and diplopia may occur. The body temperature can sometimes be high but usually does not draw attention. The infection is not limited to the meninges, also reaching the cerebral cortex, brainstem, and cerebellum. Meningeal signs, photophobia, cranial nerve palsy (III, IV, VI, VII), and further mental confusion and progressive cranial hypertension are observed as the disease progresses. These features may last from days (acute forms), 2–4 weeks (more common, sub-acute or chronic), to months (chronic), and can be misinterpreted in the initial forms as a viral or bacterial meningitis, as a chronic form of tuberculous meningitis, or as tumors. Compression of the brainstem, herniation of the cerebellar tonsils, coma, and opisthotonus are final complications that can lead to death (Diamond, 2000; Kwon-Chung & Bennett, 1992). The symptoms resulting from CNS infection are more evident and premature in patients with no evidence of immunosuppression, whereas in those with severe immunosuppression, clinical symptoms are poor and the diagnosis depends exclusively on a good laboratory routine.

Meningoencephalitis represents an extra-pulmonary dissemination of the fungus from the respiratory tract to the CNS. In the beginning or even in the later stages of the infection, neck stiffness is not observed in most cases, and cerebrospinal fluid (CSF) may present normal cellularity, mainly in immunosuppressed hosts such as those patients living with HIV. Cryptococcal meningitis may be asymptomatic or with general complaints as weight loss, fatigue, or fever. Due to these unspecific symptoms, lumbar puncture is frequently delayed, especially in resource-limited settings.

Disseminated cryptococcosis can reach any organ or system such as skin, the osteoarticular system, prostate, bone marrow, and eyes. Cryptococcal fungemia may present different skin lesions such as acneiform, papules, vesicles, tumors, abscesses, granulomas, plaques, cellulitis, or ulcerations, with drainage of pus usually rich in fungal content (Ghigliotti et al., 1992).

Primary skin lesion by traumatic inoculum is rare and may occur due to laboratory accidents and direct trauma (Neuville et al., 2003). Cryptococcemia or cryptococcal fungemia may present with fever, tremors, and chills in individuals with high fungal burden that can be revealed by blood culture. Cryptococcal peritonitis occurs in individuals undergoing peritoneal dialysis or with hepatic cirrhosis. Other sites can also be infected, although less frequently, such as the subcutaneous tissue, muscle, heart, adrenal gland, thyroid, gastrointestinal tract, and lymph nodes. Infection in the genitourinary tract is usually asymptomatic, but pyelonephritis may occur, especially in diabetic patients. The elimination of the fungus by the urine is frequently observed in cases of fungemia and can be detected by culture exam (Pinto-Junior et al., 2006).

4.3 Laboratory diagnosis

The diagnosis of cryptococcosis is traditionally based on fungal isolation from clinical specimens and/or visualization of fungal cells by direct exam or histopathology of tissues with specific stains to identify the capsule (mucicarmine) or the melanin (Fontana-Masson) (Kwon-Chung & Bennett, 1992). The development of a point-of-care test (lateral flow immunoassay; LFA) for the detection of capsular antigen (CrAg) in CSF and blood is a major advance. It requires minimal infrastructure and training, and provides a result within 10 minutes. The critical factor for best outcome will be the clinician's risk assessment and consideration of cryptococcal meningitis as a possible diagnosis during early symptoms, and then to apply the best diagnostic strategies.

Figure 4.1   *Cryptococcus neoformans* cells identified by direct microscopy (×400). India ink preparation of the cerebrospinal fluid (CSF).

4.3.1 Direct exam

The most important exam for the diagnosis of meningoencephalitis is the CSF analysis, which must be centrifuged before the laboratory procedures. Direct exam of CSF with India ink is a powerful and fast method for meningoencephalitis diagnosis, revealing the polysaccharide capsule (Figure 4.1). The test is positive for over 80% of AIDS patients, as most of them have a high fungal burden at the time of diagnosis.

The main virulence factors of *C. neoformans/C. gattii* species complexes include important phenotypic characteristics for diagnosis, such as capsule production, melanin formation, and thermotolerance (Kwon-Chung & Bennett, 1992).

4.3.2 Capsular antigen detection

Early diagnosis is the key to an effective treatment, particularly in patients in resource-limited settings. The lateral flow immunoassay for cryptococcal antigen allows a rapid and inexpensive diagnosis of cryptococcosis at or near the point of patient contact. It is based on the detection of glucuroxylomannan antigen from the fungal capsule by monoclonal antibodies impregnated in a test strip. CrAg-LFA assays present a higher sensitivity and specificity than the conventional latex agglutination, are less dependent on technician skill, and are easier to perform (Limper et al., 2017). Moreover, the pre-emptive treatment for isolated antigenemia prevents the onset of meningoencephalitis, especially when the India ink exam is negative or not performed. Screening for cryptococcal antigen (serum, plasma, or whole-blood finger-prick sample) in patients with less than 100 cells/mm³ CD4 counts is cost effective in countries with prevalence of antigenemia above 3%, thus enabling earlier diagnosis and proper treatment in those patients (Limper et al., 2017).

Cryptococcal meningitis diagnosis usually relies on lumbar puncture. The CSF with India ink preparation is positive in 70%–90% of patients with HIV-associated infection (Limper et al., 2017) because the fungal load is usually high. However, in immunocompetent hosts, as well as in those at the early stage of meningitis, direct examination may be negative. Therefore, a negative CSF direct exam requires cryptococcal antigen detection, which will reliably diagnose the infection.

The CrAg test has no diagnostic value in cases with a prior history of cryptococcosis, since the antigen can remain for a long period in the blood system, up to years. Therefore, relapses can only be diagnosed on a microbiological basis, that is, by isolation of the agent from clinical specimens.

4.3.3 Isolation and identification of
C. neoformans and *C. gattii* species

The culture of clinical specimens must be on Sabouraud 2% glucose at 25°C for at least 4 weeks incubation. Mycosel or other media with cycloheximide inhibit *Cryptococcus* spp. growth. The culture of clinical specimens in which other microorganisms of the normal biota can be found, like sputum, urine, and skin, should be done on a medium with phenolic compounds to stimulate *Cryptococcus* spp. to produce melanin, like niger seed agar (NSA), sunflower seed, caffeic acid, or L-Dopa media (Figure 4.2). Thus, *C. neoformans* and *C. gattii* stand out from other yeasts by forming dark brown colonies, making their purification easier for further species identification.

C. neoformans and *C. gattii* present micro morphology and a set of common physiological and biochemical characteristics. The main characteristics are polysaccharide capsule, no hypha or

Figure 4.2 Dark brown melanized colonies of *Cryptococcus neoformans* grown on niger seed agar (NSA).

pseudohypha, thermotolerance at 37°C, melanin production on NSA medium (or similar media), and urease production (it is a rare urease-negative variant). In addition, they are non-fermenters, and assimilate the following compounds by oxidative metabolism: galactose, sucrose, maltose, trehalose, melizitose, D-xylose, L-rhamnose, sorbitol, mannitol, dulcitol, D-mannitol, α-methyl-d-glucoside, salicin, inositol, glucose, and fructose. They do not assimilate nitrate and are sensitive to cycloheximide, not being able to grow on media containing this drug at the concentrations of 0.2% to 0.5%, although some isolates may grow at lower concentrations (Kwon-Chung & Bennett, 1992).

As the commercial tests for assimilation of carbon and nitrogen sources do not distinguish *C. gattii* from *C. neoformans*, these species can be differentiated using the selective medium l-canavanine glycine bromothymol blue (CGB) agar, produced in house or using the commercial CGB agar (Thermo Fisher Scientific Inc., USA). The blue color of glycine assimilation and resistance to L-canavanine on CGB agar indicates the positive reaction caused by *C. gattii*, whereas *C. neoformans* fails to cause a color change (Min & Kwon-Chung, 1986).

Molecular identification is not routinely applied to the diagnosis of cryptococcosis and demands the complete sequence of the Internal Transcribed Spacer (ITS) and/or partial sequence of the nuclear ribosomal RNA gene large subunit (D1–D2 domains of 26/28S) (Stielow et al., 2015).

Recently, the technology of matrix-assisted laser desorption/ionization time-of-flight mass spectrometry (MALDI-TOF MS) has been used in the identification of yeasts as a substitute for phenotypic and genotypic methodologies. This technique determines within a few minutes specific patterns of peptides and protein mass spectra of either intact microbial cells or cellular extracts from pure cultures or from biological samples, allowing a reliable species- and subspecies-level identification by comparison of the obtained spectrum with those included in a reference spectral library. Thus, it is dependent on database quality (Firacative et al., 2012).

4.3.4 Histopathology

Histopathologic identification in biopsy specimens is based on the micromorphological characteristics of *Cryptococcus* spp. Hematoxylin-eosin stain shows the pattern of host response and can reveal the fungal structures. Gomori-Grocott's silver stain reveals the fungal wall, but does not discriminate the capsule. The Fontana-Masson stain evidences melanin deposition, contributing to the diagnosis. The most important stain for cryptococcosis diagnosis is Mayer's Mucicarmin, which blends the polysaccharide capsule in red. However, none of these techniques can differentiate the species *C. neoformans* from *C. gattii*.

4.4 Therapy

The combination of amphotericin B deoxycholate (D-AmB; 0·7–1·0 mg/kg per day) and flucytosine (100 mg/kg per day in four divided doses) for the initial 2 weeks, followed by fluconazole (400–800 mg per day for 8 weeks, and 200 mg per day thereafter) is the gold standard for antifungal treatment of cryptococcal meningitis in HIV and non-HIV, as well as for other disseminated forms. The addition of flucytosine is associated with 40% reduction in mortality compared with D-AmB alone (Limper et al., 2017; Williamson et al., 2017).

The most frequent complication of cryptococcal meningitis is raised cerebrospinal fluid pressure that is associated with increased mortality. Careful therapeutic lumbar punctures are effective and only severely ill patients might require a temporary lumbar drain or ventricular shunt. Paradoxical cryptococcal immune reconstitution inflammatory syndrome (IRIS) is observed in 15%–20% of HIV patients treated for cryptococcal meningitis, usually observed around 1 month after starting the antiretroviral therapy (ART). Short courses of corticosteroids have been used successfully, but corticosteroids given with initial antifungal therapy are harmful (Limper et al., 2017).

Antifungal therapy for pulmonary cryptococcosis without clinical and laboratorial evidence of dissemination is based on oral fluconazole or itraconazole, 200–400 mg/day during 6–12 months.

4.5 Prevention

Pre-emptive fluconazole treatment strategy to prevent the development of meningitis in patients with low CD4 cell counts (lower than 100 cells per µL) and serum CrAg positive has been endorsed in the WHO guidelines. Such a strategy can be highly cost effective, and screening has been introduced in South Africa and elsewhere (Limper et al., 2017).

References

Acheson, E. S., Galanis, E., Bartlett, K. et al. 2018. Searching for clues for eighteen years: Deciphering the ecological determinants of *Cryptococcus gattii* on Vancouver Island, British Columbia. *Med Mycol* 56: 129–144.

Alves, G. S., Freire, A. K., Bentes, A. S. et al. 2016. Molecular typing of environmental *Cryptococcus neoformans/C. gattii* species complex isolates from Manaus, Amazonas, Brazil. *Mycoses* 59: 509–515.

Brito-Santos, F., Barbosa, G. G., Trilles, L. et al. 2015. Environmental isolation of *Cryptococcus gattii* VGII from indoor dust from typical wooden houses in the deep Amazonas of the Rio Negro basin. *PLOS ONE* 10: e0115866.

Chen, Y. H., Yu, F., Bian, Z. Y. et al. 2018. Multilocus sequence typing reveals both shared and unique genotypes of *Cryptococcus neoformans* in Jiangxi province, China. *Sci Rep* 8: 1495.

Diamond, R. D. 2000. Cryptococcus neoformans. In *Principles and Practice of Infectious Diseases*, (eds). G. L. Mandell, J. E. Bennet, and R. Dollin, 2331–2340, 4th ed., Churchill Livinsgstone, Pennsylvania.

Firacative, C., Trilles, L., Meyer, W. 2012. MALDI-TOF MS enables the rapid identification of the major molecular types within the *Cryptococcus neoformans/C. gattii* species complex. *PLOS ONE* 7: e37566.

Ghigliotti, G., Carrega, G., Farris, A. et al. 1992. Cutaneous cryptococcosis resembling molluscum contagiosum in a homosexual man with AIDS. Report of a case and review of the literature. *Acta Derm Venereol* 72: 182–184.

Kim, Y. S., Lee, I. H., Kim, H. S. et al. 2012. Pulmonary cryptococcosis mimicking primary lung cancer with multiple lung metastases. *Tuberc Respir Dis (Seoul)* 73: 182–186.

Kwon-Chung, K. J., Bennett, J. E. 1992. Cryptococcosis. In *Medical Mycology*. Lea & Febiger, Philadelphia.

Kwon-Chung, K. J., Bennett, J. E., Wickes, B. L. et al. 2017. The case for adopting the "Species Complex. *mSphere* 2(1): e00357–16.

Kwon-Chung, K. J., Fraser, J. A., Doering, T. L. et al. 2014. *Cryptococcus neoformans* and *Cryptococcus gattii*, the etiologic agents of cryptococcosis. *Cold Spring Harb Perspect Med* 4: a019760.

Limper, A. H., Adenis, A., Le, T., Harrison, T. S. 2017. Fungal infections in HIV/AIDS. *Lancet Infect Dis* 17: e334–e343.

Lockhart, S. R., Iqbal, N., Harris, J. R. et al. 2013. *Cryptococcus gattii* in the United States: Genotypic diversity of human and veterinary isolates. *PLOS ONE* 8: e74737.

Martins, L. M., Wanke, B., Lazéra, M. S. et al. 2011. Genotypes of *Cryptococcus neoformans* and *Cryptococcus gattii* as agents of endemic cryptococcosis in Teresina, Piauí (northeastern Brazil). *Mem Inst Oswaldo Cruz* 106: 725–730.

Maziarz, E. K., Perfect, J. R. 2016. Cryptococcosis. *Infect Dis Clin North Am.* 30: 179–206.

Min, K. H., Kwon-Chung, K. J. 1986. The biochemical basis for the distinction between the two *Cryptococcus neoformans* varieties with CGB medium. *Zentralbl Bakteriol Mikrobiol Hyg A* 261: 471–480.

Neuville, S., Dromer, F., Morin, O. et al. 2003. Primary cutaneous cryptococcosis: A distinct clinical entity. *Clin Infect Dis* 36: 337–347.

O'Halloran, J. A., Powderly, W. G., Spec, A. 2017. Cryptococcosis today: It is not all about HIV infection. *Curr Clin Micro Rep* 4: 88–95.

Pinto-Junior, V. L., Galhardo, M. C., Lazéra, M. et al. 2006. Importance of culture of urine in the diagnosis of AIDS associated cryptococcosis. *Rev Soc Bras Med Trop* 39: 230–232.

Rajasingham, R., Smith, R. M., Park, B. J. et al. 2017. Global burden of disease of HIV-associated cryptococcal meningitis: An updated analysis. *Lancet Infect Dis* 17(8): 873–881.

Santos, W. R., Meyer, W., Wanke, B. et al. 2008. Primary endemic *Cryptococcosis gattii* by molecular type VGII in the state of Pará, Brazil. *Mem Inst Oswaldo Cruz* 103: 813–818.

Stielow, J. B., Lévesque, C. A., Seifert, K. A. et al. 2015. One fungus, which genes? Development and assessment of universal primers for potential secondary fungal DNA barcodes. *Persoonia* 35: 242–263.

Trilles, L., Lazera, M., Wanke, B. et al. 2003. Genetic characterization of environmental isolates of the Cryptococcus neoformans species complex from Brazil. **Med Mycol** 41: 383–390.

Trilles, L., Lazéra, M. S., Wanke, B. et al. 2008. Regional pattern of the molecular types of *Cryptococcus neoformans* and *Cryptococcus gattii* in Brazil. *Mem Inst Oswaldo Cruz* 103: 455–462.

Williamson, P. R., Jarvis, J. N., Panackal, A. A. et al. 2017. Cryptococcal meningitis: Epidemiology, immunology, diagnosis and therapy. *Nat Rev Neurol* 13: 13–24.

5
Trichosporon

João Nobrega de Almeida Júnior

Contents

5.1 Introduction

The *Trichosporon* species are basidiomycetous yeast-like organisms that are ubiquitous and found predominantly in warm climate regions (Colombo et al., 2011). These organisms can be found in soil, decomposing wood, air, water, mammals, insects, and hospital environments (Colombo et al., 2011; Fanfair et al., 2013). In humans, some *Trichosporon* species, like *Trichosporon asahii*, may be part of the gastrointestinal and skin microbiota (Cho et al., 2015; Ellner et al., 1991). However, *Trichosporon* has been also related to human diseases, such as white piedra, summer-type hypersensitivity pneumonitis (SHP), and deep-seated infections, such as fungemia, in patients with immunodepression or invasive devices (Colombo et al., 2011; De Almeida Júnior & Hennequin, 2016).

Until 30 years ago, *Trichosporon beigelii* was considered the only species of the genus. However, a taxonomic revision conducted by Guého et al. based on molecular analysis demonstrated that isolates considered *T. beigelii* had high phylogenetic diversity, which led them to describe other species such as *T. asahii*, *Trichosporon asteroides*, *Trichosporon cutaneum*, *Trichosporon inkin*, *Trichosporon mucoides*, and *Trichosporon ovoides* (Guého et al., 1992). Over the years, several *Trichosporon* species have been described, accounting for a total of 50 different taxa divided into four clades, named Ovoides, Porosum, Cutaneum, Gracile/Brassicae (Colombo et al., 2011). Among the described species, 18 have been recovered in clinical samples, including *T. asahii*, *T. inkin*, *T. asteroides*, *T. ovoides*, *Trichosporon faecale*, *Trichosporon coremiiforme*, *Trichosporon japonicum*, *Trichosporon lactis*, and *Trichosporon dohaense* from the clade Porosum; *T. cutaneum*, *Trichosporon dermatis*, *Trichosporon mucoides*, *Trichosporon jirovecii*, and *Trichosporon debeurmannianum* from the clade Cutaneum; and *Trichosporon montevideense*, *Trichosporon domesticum*, *Trichosporon loubieri*, and *Trichosporon mycotoxinivorans* from the clade Gracile/Brassicae (Almeida et al., 2017; Colombo et al., 2011; De Almeida et al., 2017). However, in 2015, a new taxonomic classification, based on multiple-gene sequence analyses, split the genus *Trichosporon* into three genera. The species belonging to the former clade Gracile/Brassicae now belong to the genus *Apiotrichum*, while the species belonging to the former clade Cutaneum are now the genus *Cutaneotrichosporon* (Liu et al., 2015).

The identification of *Trichosporon* in clinical samples has epidemiological and clinical relevance. White piedra is frequently misdiagnosed as pediculosis, and correct diagnosis avoids unnecessary and toxic treatments (Fischman et al., 2014). *Trichosporon* has been related to nosocomial infection outbreaks, and rapid identification may hasten the implementation of control measures (Fanfair et al., 2013). And, finally, *Trichosporon* is an emerging pathogen of deep-seated infections, and fast and accurate diagnosis may accelerate the prescription of the best effective therapy and ameliorate the patients' prognosis (De Almeida Júnior & Hennequin, 2016).

5.2 Specimen selection, collection, and transport

For the diagnosis of white piedra, cut off hair samples with nodules attached. Samples can be directly inoculated on Sabouraud dextrose agar (SDA, Difco, USA) with chloramphenicol (50 mg/L), or put between two clean glass slides, taped together, and sent to the mycology laboratory in a slide carrier. Both microscopic examination and culture must be requested (Jorgensen et al., 2015).

In case of deep-seated *Trichosporon* infection, sample collection must follow the general guidelines for bacterial and fungal infections since *Trichosporon* may show growth in common bacterial (i.e., blood sheep agar) and fungal media (i.e., SDA) (Jorgensen et al., 2015). Samples must be collected aseptically, before the introduction of antifungals, placed in sterile leak-proof containers, and delivered to the laboratory within 2 hours at room temperature.

Patients with disseminated infections may develop fungemia and pulmonary and skin lesions. Two or three sets of blood cultures must be collected for the diagnosis of bloodstream infection, with 8–10 mL of blood in each bottle. *T. asahii* grows better on aerobic standard blood culture bottles than anaerobic ones, and bottles are flagged positive over 30 hours after incubation on automated incubators (Nawrot et al., 2015). *Trichosporon* usually colonizes central venous catheters (CVCs), which may be the source of a bloodstream infection episode. Thus, catheter tips or blood cultures from the CVC lumen may be useful for the diagnosis of a CVC-related bloodstream infection.

Bronchoalveolar lavage or tracheal aspirates may be useful for the diagnosis of *Trichosporon* pneumonia. Colombo et al. proposed the criteria for probable *Trichosporon* pneumonia in at-risk patients (i.e., neutropenia; persistent fever despite antimicrobial therapy): pulmonary infiltrates and recovery of *Trichosporon* in respiratory samples in the absence of other pathogens causing opportunistic infections (Colombo et al., 2011). Finally, lung or skin biopsies may be useful for the diagnosis of disseminated *Trichosporon* infection. They must be split for histopathological and microbiological analyses.

5.3 Specimen processing in the mycology laboratory

Hair samples with nodules must have a microscopic examination on 20% potassium hydroxide mounts, inoculated on Sabouraud dextrose agar with antibiotics (e.g., chloramphenicol [50 mg/L]) and incubated at 30°C for at least 30 days (Jorgensen et al., 2015).

Biopsies must have a carefully microscopic examination on 20% potassium hydroxide. Fluorescent brighteners like calcofluor white help to better visualize fungal elements among tissue cells and components. For fungal cultures, the biopsies should be minced and in several pieces and inoculated on different media, including selective and non-selective agar, selective broth (e.g., brain heart infusion with antibiotics such as gentamicin and chloramphenicol), and incubated at 30°C for 4 weeks according to the standard protocols (Jorgensen et al., 2015). For inoculation of solid media, previous-scratch the agar and place the tissue into the scratch for better recovery of the fungi.

Respiratory samples must be inoculated on SDA with antibiotics and incubated at 30°C for at least 4 weeks according to the standard protocols (Jorgensen et al., 2015). Samples inoculated on standard bacterial culture media (MacConkey, blood sheep, and chocolate agar) and on *Burkholderia cepacia* selective agar should be carefully examined since *Trichosporon* colonies may also show growth on those bacteriological media.

Positive blood cultures with yeasts by Gram stain must be inoculated on SDA and incubated for at least 48 hours at 30°C. Seeding on chromogenic agar may be helpful to separate colonies when there is a polymicrobial growth. CVC tips are inoculated on blood sheep agar using the standard rolling plate method and incubated at 35°C for 48 hours in a 5% CO_2 atmosphere (Jorgensen et al., 2015). Any yeast growth must be identified and reported. There is no cutoff value to discriminate colonization from infection in CVC-related *Trichosporon* fungemia.

5.4 Species identification

Microscopic examination may provide useful information for the diagnosis of a *Trichosporon* infection. The presence of blastoconidia and arthroconidia are very suggestive (Figure 5.1A). On SDA media, colonies are usually cream-colored, dry (exception *Trichosporon dermatis/ mucoides*), farinose, and become cerebriform after 7 days at room temperature (Figure 5.2). On chromogenic media, *T. asahii* colonies develop a blue-dry aspect (Figure 5.3A).

To differentiate *Trichosporon* from the ascomycete *Geotrichum*, the hydrolysis of urea is a useful test (Figure 5.3B). This simple test, allied to macro- and micromorphology, provides genus identification. A slide culture on cornmeal agar medium with Tween 20 allied with other phenotypic tests may provide useful information for species identification (Figure 5.1B, Table 5.1). However, closely related species may show identical phenotypic profiles and species misidentifications may occur. Commercial manuals (e.g., *API 20 AUX* [bioMérieux], AuxaColor 2 [Bio-Rad]) or automated tests (Vitek 2; bioMérieux) claim to provide identification of *T. asahii, T. inkin*, and *T. mucoides*. However, misidentifications of *T. inkin* as *T. asahii*, and *T. dermatis* as *T. mucoides* are common (De Almeida Júnior et al., 2014; De Figueiredo et al., 2014), and other species may lack identification at the genus level (De Figueiredo et al., 2014).

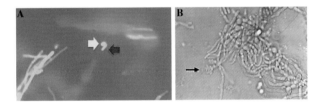

Figure 5.1   Micromorphology of *Trichosporon*. (A) Hyphae, blastoconidium (yellow arrow) and arthroconidium (red arrow) in a skin biopsy from a patient with acute myeloid leukemia (calcofluor white stain, 400×). (B) Slide culture showing apressoria (black arrow) of a *T. inkin* isolate. (Adapted from Elaine C. Francisco, Special Medical Mycology Laboratory, Federal University of São Paulo.)

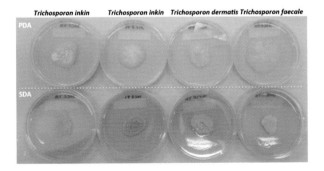

Figure 5.2   *Trichosporon* colonies on potato dextrose agar (PDA), Sabouraud dextrose agar (SDA), after 3 days at 30°C. (Adapted from Dulce Sachiko Yamamoto de Figueiredo, Laboratory of Medical Mycology-LIM53, Instituto de Medicina Tropical, Federal University of São Paulo.)

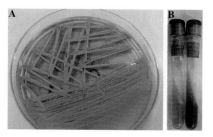

Figure 5.3   Presumptive diagnosis of Trichosporon (A) *T. asahii* colonies on ChromoCandida medium after 48 hours incubation at 37°C. (B) Urea hydrolysis test. Basidiomycetes produce urease, which hydrolyzes urea to ammonia and carbon dioxide. The formation of ammonia alkalinizes the medium (right tube), and the pH shift is detected by the color change of phenol red from light orange at pH 6.8 to pink at pH 8.1. (Adapted from [a] Central Laboratory Division, Hospital das Clínicas, Faculty of Medicine, Federal University of São Paulo; [b] Afonso Rafael da Silva Jr., Central Laboratory Division, Hospital das Clínicas, Faculty of Medicine, Federal University of São Paulo.)

Sequence analysis of the intergenic spacer from the ribosomal DNA with the primers 26S (5'-ATCCTTTGCAGACGACTTGA-3') and 5SR (5'-AGCTTGACTTCGCAGATCGG-3') is the gold-standard method for species identification (Sugita et al., 2002). Some closely related species like *T. asteroides* and *T. japonicum* are only differentiated by a single nucleotide polymorphism (Sugita et al., 2002).

MALDI-TOF mass spectrometry has emerged as a revolutionary method for microorganism identification in the microbiology laboratory. This technique has shown comparable results to rDNA sequence analysis for *Trichosporon* and closely related genera species identification (De Almeida et al., 2017; De Almeida Júnior et al., 2014; Kolecka et al., 2013). Moreover, MALDI-TOF MS is less laborious, faster, and cheaper than molecular methods and, with the improvement of the reference spectra libraries, it might be, in the near future, the reference technique for *Trichosporon* species identification in clinical laboratories.

5.5 Rapid diagnostic tests

Sera from patients with disseminated *Trichosporon* infection may show cross-reactivity with assays design to detect the capsular antigen glucuronoxylomannan (GXM) from *Cryptococcus* (Mekha et al., 2007; Rivet-Dañon et al., 2015). It has been demonstrated that *Trichosporon asahii* also have GXM anchored to the cell wall (Fonseca et al., 2009), which explains the aforementioned cross-reactivity. Thus, patients with risk factors for invasive *Trichosporon* infection may have a positive serum GXM test early in the course of the disease.

An in-house real-time PCR assay designed with *T. asahii*-specific primers/probe has been shown to detect *T. asahii* DNA in blood samples of trichosporonosis patients, with higher sensitivity than the polysaccharide antigen assay for *Cryptococcus* (Mekha et al., 2007).

Direct identification of *Trichosporon* from positive blood cultures by MALDI-TOF MS using a lysing-centrifugation extraction process is an alternative and promising method for a faster diagnosis of *Trichosporon* fungemia; however, it still needs multicenter validation (De Almeida et al., 2016).

5.6 Serology

Trichosporon serology has been useful for the diagnosis of SHP. Specific anti-*Trichosporon* antibodies are usually detected by enzyme immunoassays, although commercial kits are not available outside Japan (Mizobe et al., 2002).

Table 5.1 Phenotypic Aspects That Might be Helpful to Discriminate the Main *Trichosporon* Pathogenic Species

Species	Growth at 30°C	Growth at 37°C	Growth at 40°C	Growth at 42°C	Apressoria on 2% MEA	Meristematic Conidia on 2% MEAr	Sensitivity to 0.1% Cyclohexemide	Nitrite
T. asahii	+	+	V	–	–	–	–	V
T. asteroids	+	+	V	–	–	+	–	–
T. coremiiforme	+	V	V	–	–	–	+	+
T. faecale	+	+	+	–	–	–	–	–
T. ovoides	+	V	–	–	+	–	–	V
T. inkin	+	+	+	+	+	–	+	+
T. dermatis[a]	+	+	–	–	–	–	–	+
T. mucoides[a]	+	+	–	–	–	–	–	+
T. mycotoxinivorans[b]	+	+	–	–	–	–	–	–

Source: Kutzman, C. et al. 2011. *The Yeasts, a Taxonomic Study.* 5th edition, Elsevier Science.
Abbreviations: MEA = malt extract agar; V = variable.
[a] According to a new taxonomic classification (Liu et al., 2015), the new genus *Cutaneotrichosporon* has been proposed for this species.
[b] According to a new taxonomic classification (Liu et al., 2015), the new genus *Apiotrichum* has been proposed for this species.

References

Almeida, J. N. de, Francisco, E. C., Barberino, M. G. M. de A. et al. 2017. Emergence of *Trichosporon mycotoxinivorans* (*Apiotrichum mycotoxinivorans*) invasive infections in Latin America. *Mem Inst Oswaldo Cruz* 112: 719–722.

Cho, O., Matsukura, M., Sugita, T. 2015. Molecular evidence that the opportunistic fungal pathogen *Trichosporon asahii* is part of the normal fungal microbiota of the human gut based on rRNA genotyping. *Int J Infect Dis* 39: 87–88.

Colombo, A. L., Padovan, A. C. B., Chaves, G. M. 2011. Current knowledge of *Trichosporon* spp. and Trichosporonosis. *Clin Microbiol Rev* 24: 682–700.

De Almeida, J. N., Favero Gimenes, V. M., Francisco, E. C. et al. 2017. Evaluating and improving Vitek MS for identification of clinically relevant species of *Trichosporon* and the closely related genera *Cutaneotrichosporon* and *Apiotrichum*. *J Clin Microbiol* 55: 2439–2444.

De Almeida, J. N., Sztajnbok, J., da Silva, A. R. et al. 2016. Rapid identification of moulds and arthroconidial yeasts from positive blood cultures by MALDI-TOF mass spectrometry. *Med Mycol* 54: 885–889.

De Almeida Júnior, J. N., Figueiredo, D. S. Y., Toubas, D., Del Negro, G. M. B., Motta, A. L., Rossi. F. 2014. Usefulness of MALDI-TOF mass spectrometry for identifying clinical *Trichosporon* isolates. *Clin Microbiol Infect* 20: 784–790.

De Almeida Júnior, J. N., Hennequin, C. 2016. Invasive *Trichosporon* infection: A systematic review on a re-emerging fungal pathogen. *Front Microbiol* 7: 1629.

De Figueiredo, D. S. Y., de Almeida, J. N., Motta, A. L., Castro E Silva, D. M., Szeszs, M. W., Del Negro, G. M. B. 2014. Evaluation of VITEK 2 for discriminating Trichosporon species: Misidentification of *Trichosporon* non-*T. asahii*. *Diagn Microbiol Infect Dis* 80: 59–61.

Ellner, K., McBride, M. E., Rosen, T., Berman, D. 1991. Prevalence of *Trichosporon beigelii*. Colonization of normal perigenital skin. *J Med Vet* 29: 99–103.

Fanfair, R. N., Heslop, O., Etienne, K. et al. 2013. *Trichosporon asahii* among intensive care unit patients at a medical center in Jamaica. *Infect Control Hosp Epidemiol* 34: 638–641.

Fischman, O., Bezerra, F. C., Francisco, E. C. et al. 2014. *Trichosporon inkin*: An uncommon agent of scalp white piedra. Report of four cases in Brazilian children. *Mycopathologia* 178: 85–89.

Fonseca, F. L., Frases, S., Casadevall, A., Fischman-Gompertz, O., Nimrichter, L., Rodrigues, M. L. 2009. Structural and functional properties of the *Trichosporon asahii* glucuronoxylomannan. *Fungal Genet Biol* 46: 496–505.

Guého, E., Smith, M. T., de Hoog, G. S., Billon-Grand, G., Christen, R., Batenburg-van der Vegte, W. H. 1992. Contributions to a revision of the genus *Trichosporon*. *Antonie Van Leeuwenhoek* 61: 289–316.

Jorgensen, J., Pfaller, M., Carroll, K., Funke, G., Landry, M., Richter, S., Warnock D. (eds). 2015. *Manual of Clinical Microbiology*. 11th edition. ASM Press, Washington, DC.

Kolecka, A., Khayhan, K., Groenewald, M. et al. 2013. Identification of medically relevant species of arthroconidial yeasts by use of matrix-assisted laser desorption ionization–time of flight mass spectrometry. *J Clin Microbiol* 51: 2491–2500.

Kutzman, C., Fell, J. W., Boekhout, T. 2011. *The Yeasts, a Taxonomic Study*. 5th edition, Elsevier Science.

Liu, X. Z., Wang, Q. M., Theelen, B., Groenewald, M., Bai, F. Y., Boekhout, T. 2015. Phylogeny of tremellomycetous yeasts and related dimorphic and filamentous basidiomycetes reconstructed from multiple gene sequence analyses. *Stud Mycol* 81: 1–26.

Mekha, N., Sugita, T., Ikeda, R., Nishikawa, A., Poonwan, N. 2007. Real-time PCR assay to detect DNA in sera for the diagnosis of deep-seated trichosporonosis. *Microbiol Immunol* 51: 633–635.

Mizobe, T., Adachi, S., Hamaoka, A., Andoh, M. 2002. Evaluation of the enzyme-linked immunosorbent assay system for sero-diagnosis of summer-type hypersensitivity pneumonitis. *Arerugi* 51: 20–23.

Nawrot, U., Kowalska-Krochmal, B., Sulik-Tyszka, B. et al. 2015. Evaluation of blood culture media for the detection of fungi. *Eur J Clin Microbiol Infect Dis* 34: 161–167.

Rivet-Dañon, D., Guitard, J., Grenouillet, F. et al. 2015. Rapid diagnosis of cryptococcosis using an antigen detection immunochromatographic test. *J Infect* 70: 499–503.

Sugita, T., Nakajima, M., Ikeda, R., Matsushima, T., Shinoda, T. 2002. Sequence analysis of the ribosomal DNA intergenic spacer 1 regions of *Trichosporon* species. *J Clin Microbiol* 40: 1826–1830.

6

Malassezia

Reginaldo Gonçalves de Lima-Neto, Danielle Patrícia
Cerqueira Macêdo, Ana Maria Rabelo de Carvalho,
Carolina Maria da Silva, and Rejane Pereira Neves

Contents

6.1 Introduction

Members of the genus *Malassezia* spp. are part of human and animal microbiota, belonging to the phylum Basidiomycota, class Malasseziomycetes. Under certain circumstances, these fungi are potentially pathogenic to humans and domestic animals. They commonly cause superficial lesions, but in rare cases also cause systemic diseases (Ashbee and Evans, 2002; Talaee et al., 2014). Recently, several reports have indicated that *Malassezia* antigens are important triggers for the development and/or exacerbation of atopic dermatitis and psoriasis (Cafarchia et al., 2011; Prohić, 2012).

Since the first description of *M. furfur* (Robin) Bailon 1889, the taxonomy of the genus has been a matter of controversy. Advances in molecular biology have allowed the description of 16 species, 6 of them found exclusively in animals (Table 6.1). For the majority of *Malassezia* species, however, details regarding physiology, biochemistry, and pathogenesis are still lacking (Cafarchia et al., 2011; Gaitanis et al., 2012; Prohić, 2012; Cabañes et al., 2016; Prohić et al., 2016).

Malassezia are lipophilic or lipid-dependent species and their primary isolation from clinical samples and cultivation are challenging because these fungi do not grow on conventional mycological media. In addition, they are not easily identified by conventional phenotypic methods, so a molecular approach is necessary for definitive diagnosis. However, routine laboratories must make efforts to establish a routine for proper identification of *Malassezia* isolates, at least those obtained from deep infections.

6.2 Pathophysiology of superficial infections

Malassezia species play a key role in the stability of the normal skin microbiota and in maintaining health (Peleg et al., 2010). These commensal fungi colonize the host from the first weeks of life (Marples, 1965; Findley et al., 2013). In healthy skin, *Malassezia* species generally use nutrients essential for their growth without causing disease. However, this relationship can be altered if

Table 6.1   Origin of Isolation of *Malassezia* Species

Species	Species Description	Isolation from Humans		Animal Species
		Healthy	Diseased	
M. furfur	Bailon (1889)	+	AD; SD; PS; FL; PV; onychomycosis; systemic infections	Domestic animals: healthy skin
M. pachydermatis	(Weidman) Dodge (1925)	−	Systemic infections	Domestic animals (cats, dogs): healthy skin; SD
M. sympodialis	Simmons and Guého (1990)	+	AD; SD; PS; FL; PV; otitis externa; onychomycosis; systemic infections	Sheep, horses: healthy skin Cats: otitis
M. obtusa	Guého et al. (1996)	+	AD; SD; PS; PV	−
M. globosa	Guého et al. (1996)	+	AD; SD; PS; FL; PV; onychomycosis	Cow: healthy skin Otitis in cats
M. slooffiae	Guého et al. (1996)	+	AD; SD; PS; PV; systemic infections	Pigs: healthy skin Goats: SD
M. restricta	Guého et al. (1996)	+	AD; SD; PS; FL; PV; onychomycosis; deep infection	−
M. dermatis	Sugita et al. (2002)	+	AD; SD; PS; FL; PV	−
M. japonica	Sugita et al. (2003)	+	AD; SD; PS; PV	−
M. nana	Hirai et al. (2004)	−	−	Cats, cows: otitis, skin diseases
M. yamatoensis	Sugita et al. (2004)	+	AD; SD; PS; PV	−
M. equina	Cabañes et al. 2007	−	−	Horses: ear canal
M. caprae	Cabañes et al. (2007)	−	−	Goats: healthy skin
M. cuniculi	Cabañes et al. (2011)	−	−	Rabbits: healthy skin
M. brasiliensis	Cabañes et al. (2016)	−	−	Parrots: lesions on beak
M. psittaci	Cabañes et al. (2016)	−	−	Parrots: lesions on beak

Source: Based on Cabañes F. J. et al. 2016. *Rev Iberoam Micol* 33: 92–99; Prohić A. et al. 2016. *Int J Dermatol* 55: 494–504; Harada K. et al. 2015. *J Dermatol* 42: 250–257; Prohić, A. 2012. *Psoriasis—A Systemic Disease*, InTech, Available from: https://www.intechopen.com/books/psoriasis-a-systemicdisease/psoriasis-and-malassezia-yeasts; Cafarchia C. et al. 2011. *Mol Cell Probes* 25: 1–7.
Abbreviations: AD, atopic dermatitis; SD, seborrheic dermatitis, PS, psoriasis; FL, folliculitis; PV, pityriasis versicolor.

the yeast modifies the expression of lipases and phospholipases required to obtain energy, and at the same time synthesizes malassezin and bioactive indole derivatives. These molecules bind to aryl hydrocarbon receptors expressed in epidermal cells, modifying the expression of cytokines and antimicrobial peptides and interfering with apoptosis and cell cycle (Gaitanis et al., 2008; Velegraki et al., 2015).

In patients with inflammatory skin disorders, such as atopic dermatitis, it is believed that the increase in the skin pH elicits the release of many *Malassezia* antigens in the epidermis. These molecules, and possibly the yeast cells, cross the impaired skin barrier and reach keratinocytes and dendritic cells, where they are recognized by toll-like receptors expressed by these cells, which then elicit pro-inflammatory cytokine production. *Malassezia* antigens induce specific IgE antibodies, which in turn induce production of more inflammatory molecules by mast cells (Glatz et al., 2015).

6.3 Clinical manifestations

The most frequent clinical manifestation of *Malassezia* skin infections is pityriasis versicolor— revealed by characteristic hypo- or hyperpigmented plaques with moderate or absent inflammation—due to alterations in the function of melanocytes. Pityriasis versicolor commonly affects the face and trunk, with scaling dermatosis generally being restricted to the outermost layers of the stratum corneum. The lesions may become chronic, and are more detectable during the summer. They are more common in adolescents and young adults. In some cases, pityriasis versicolor can be accompanied hyperkeratosis and acanthosis, as verified by histological examination (Hurwitz, 1981; Marcon and Powell, 1992; Lacaz et al., 2002).

Superficial *Malassezia* disorders also include onychomycosis, folliculitis, dandruff, seborrheic dermatitis, and atopic dermatitis (Carfachia et al., 2011; Harada et al., 2015; Prohić et al., 2016). Superficial lesions of the skin caused by *Malassezia* can occur at any age, but are most common in the third and fourth decades of life. These lesions can occur due to several factors, including environmental factors like high temperatures and humidity, making their occurrence more common in tropical and subtropical climates. Other factors are immunosuppression, neurological disorders, genetic predisposition, high lipid content in the skin, high serum levels of fatty acids, cortisol and androgens, poor personal hygiene practices, and stress (el-Hefnawi et al., 1971; Schmidt, 1997).

In addition, *Malassezia* may be involved in fungemia and disseminated disease (Marcon and Powell, 1992; Akaza et al., 2012). After skin colonization by *Malassezia,* many underlying conditions—parenteral nutrition with lipid supplementation, endotracheal intubation, central vascular access, necrotizing enterocolitis, intestinal perforation, abdominal surgery, invasive treatments, and use of broad-spectrum antibiotics or gastric acid inhibitors—as well as yeast-related virulence factors such as adhesion ability and biofilm formation, may predispose to opportunistic invasive infections (Gaitanis et al., 2012; Kaneko et al., 2012). In addition, the potential for dissemination of *Malassezia* can be related to immunity deficit, late withdrawal of catheters, tissue or valve injury, establishment of co-infection, and also lack or failure of antifungal treatment, leading to poor prognosis (Velegraki et al., 2015).

Catheter-associated *Malassezia* infections have been reported with increasing frequency, proving that opportunistic infections can progress to peritonitis, mixed bacterial-fungal septic arthritis, mastitis, postoperative sinusitis, fungemia, or disseminated clinical forms affecting several organs (Marcon and Powell, 1992; Lacaz et al., 2002; Arendrup et al., 2014). Following asymptomatic catheter colonization, immunosuppressed patients can present sudden respiratory deterioration (apnea), pneumonia, fever, leukopenia, leukocytosis, thrombocytopenia, vegetation of the endocardium, and other clinical signs (Marcon and Powell, 1992; Lacaz et al., 2002).

6.4 Laboratory diagnosis

6.4.1 Collection of clinical samples

The collection of clinical samples should be guided by the diagnostic hypothesis/lesion site. The Porto technique or gummed tape method is the most used for direct microscopic examination of skin lesions since it is painless and non-invasive, followed by scraping of hyper- or hypochromic lesions (Lacaz et al., 2002).

In case of suspected fungemia, blood samples should be collected aseptically and seeded in duplicate on Sabouraud dextrose agar medium supplemented with sterile olive oil, one maintained at 25°C–28°C and the other at 37°C for up to 7 days (Lacaz et al., 2002). Although culture of *Malassezia* from skin is not necessary for routine diagnosis, it can be easily performed and may be helpful for differential diagnosis in relation to other superficial mycoses (Lacaz et al., 2002; Velegraki et al., 2015).

For the laboratory diagnosis of a systemic infection by *Malassezia*, blood culture is always necessary. Blood collection should be performed by aseptic puncture, in triplicate and alternate days. However, if the infection is observed to be widespread, tissue fragments obtained by biopsy of the affected organs must be microscopically investigated and cultured (Lacaz et al., 2002; Arendrup et al., 2014; Velegraki et al., 2015; Cornu et al., 2018).

Laboratory methods to establish the diagnosis of catheter-related *Malassezia* infections are recommended for patients with clinical evidence of sepsis who are receiving lipid emulsions through a central venous catheter (CVC). In catheter-related *Malassezia* infections, direct microscopy of blood drawn through the catheter may reveal characteristic single budding yeast cells, indicating the culture is positive, but smears of peripheral blood are usually negative (Lacaz et al., 2002; Arendrup et al., 2014; Cornu et al., 2018).

6.4.2 Microscopy

The direct examination of clinical specimens by tape imprint or scrapings can reveal the presence of both yeast and hyphal forms (pleomorphism). This appearance is often referred to as "spaghetti and meatballs" when both round and linear elements are seen mixed together (Figure 6.1A). Direct mycological examination usually reveals the presence of yeast-like cells, grouped in a "grape bunch" format (Figure 6.1B). The yeast-like cells, usually having a size in the range of 1.5–4.5 μm × 3–7 μm, may show small collarettes, although these are hard to discern with a light microscope. The cells are referred to as being unipolar, where bud-like structures form singly on a broad base (Veasey et al., 2017).

6.4.3 Culture

Malassezia clinical strains are difficult to grow *in vitro,* so cultures of scrapings may be reported as negative (Kolecka et al., 2014). The yeast grows best if a lipid such as olive oil is added to Sabouraud agar. Cultivation is preferred on modified Dixon agar and Leeming and Notman agar. Some *Malassezia* colonies may grow rapidly, maturing in 5–7 days at 30°C–35°C on lipid-supplemented media. However, different *Malassezia* species show different growth rates on lipid-rich media. They have a rather narrow temperature growth range, since they grow poorly at 25°C, and some species will not grow above 37°C. Colonies are cream or beige to yellowish-brown in color (Figure 6.1C). They present a pasty aspect, often becoming wrinkled as they age. Lipid supplementation is required for growth of 15 of the 16 species of the genus. Hyphal elements are usually absent in culture, but rudimentary forms can occasionally be seen by microscopy.

6.4.4 Identification

Phenotypical characterization of *Malassezia* is a difficult task, since many species share nutritional requirements and morphological traits (Table 6.2). Currently, 16 *Malassezia* species are known, of which *M. pachydermatis* is the only lipophilic but non-lipid-dependent one. Thus, conventional

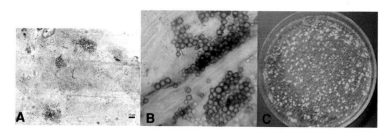

Figure 6.1   Direct mycological examination for pityriasis versicolor. (A) Cluster of yeast cells and short and tangled pseudo-hyphae with similar appearance to "spaghetti and meatballs," stained by methylene blue under gummed tape. (B) Grouped yeast-like cells in a "grape bunch" format on hair follicle stained by methylene blue under gummed tape. (C) *Malassezia* colony, yellowish-brown after 7 days of growth on Sabouraud agar culture enriched with olive oil at 25°C.

Table 6.2   Physiological Profile of *Malassezia* Species according to Cabañes et al. (2016) and Cafarchia et al. (2011)

Species	Lipid Dependence	Diffusion Tests					Catalase	β-Glucosidase	Growth in Dixon Agar at	
		Tween 20[a]	Tween 40[a]	Tween 60[a]	Tween 80[a]	Cremophor EL[b]			32°C	40°C
M. furfur	+	+[c]	+[c]	+[c]	+[c]	+[c]	+[c]	– or ±	+	+
M. pachydermatis	–	+	+	+	+	+	+	+[c]	+	+
M. sympodialis	+	– or ±	+	+	+	– or ±	+	+	+	+
M. obtusa	+	–	–	–	–	–	+	+	+	–
M. globosa	+	–	–	+	–	–	+	–	+	–
M. slooffiae	+	+ or ±	+	+	–	–	+	–	+	+
M. restricta	+	–	–	–	–	–	–	–	+	–
M. dermatis	+	+	+	+	+	+ or ±	+	–	+	+
M. japonica	+	–	±	+	–	?	+	?	+	–
M. nana	+	v	+	+	±	–	+	–	+	+ or –
M. yamatoensis	+	+	+	+	+	?	+	?	+	–
M. equine	+	+	+	+	+	–	+	–[c]	+	–
M. caprae	+	–	+	+	+[c]	–	+	+[c]	+	–
M. cuniculi	+	–	–	–	–	–	+	+	– or ±	+
M. brasiliensis	+	+	+	+	+	+	+	–	+	+
M. psittaci	+	+	+	+	+	+	+	–	+	–

Abbreviations: +, positive; –, negative; ±, weakly positive; ?, unknown results; v, variable results.
[a]   Tween diffusion test on Sabouraud agar proposed by Guillot et al. (1996).
[b]   Diffusion test proposed by Mayser et al. (1997).
[c]   Contrary results may rarely be found.

Table 6.3   User-Friendly Molecular Methods for *Malassezia* Identification in Clinical Mycological Laboratories

Molecular Tools/ Genomic Region	Species Identified	Reference
Sequencing/ITS1 rDNA	*M. furfur; M. globosa; M. pachydermatis; M. restricta; M. sympodialis; M. slooffiae; M. nana; M. obtusa*	Makimura et al. (2000); Hirai et al. (2004); Gaitanis et al. (2006); Cafarchia et al. (2008)
Sequencing/LSU rDNA	*M. furfur; M. globosa; M. restricta; M. sympodialis; M. pachydermatis; M. slooffiae; M. obtusa; M. nana*	Guillot et al. (1997); Guillot et al. (2000); Cabañes et al. (2005); Cafarchia et al. (2007, 2008)
Sequencing/LSU rDNA D1 and D2 domain/ITS	*M. furfur* (D1, D2, and ITS); *M. sympodialis* (ITS); *M. pachydermatis* (D1, D2, and ITS); *M. globosa* (D1, D2, and ITS); *M. slooffiae* (D1, D2, and ITS)	Gupta et al. (2004)
PCR-REA/ITS-1 ITS-2 rDNA and LSU rDNA	*M. furfur; M. obtusa; M. globosa; M. slooffiae; M. sympodialis; M. restricta; M. pachydermatis; M. dermatis; M. japonica; M. nana; M. yamatoensis*	Gaitanis et al. (2008); Oh et al. (2009)

Source: Data partially extracted from Cafarchia, C. et al. 2011. *Mol Cell Probes* 25: 1–7; Gaitanis, G. et al. 2012. *Clin Microbiol Rev* 25: 106–141.
Abbreviations: ITS, internal transcribed spacer (ITS1-5.8S-ITS2) of the ribosomal DNA region; REA, restriction enzyme analysis; LSU, large subunit of rDNA.

tests like Tween-based culture or evaluation of beta-glucosidase activity and growth with Cremophor EL may not be suitable to differentiate them, besides being time-consuming.

As previously mentioned, proper identification of *Malassezia* species can be attained by molecular methods. Many molecular tools have been established for *Malassezia* identification and also for the detection of intraspecific genetic variation: (a) direct sequencing of rDNA loci (internal transcribed spacers ITS-1 and ITS-2, intergenic spacer IGS-1, large subunit LSU, chitin synthase gene and RNA polymerase subunit 1 gene, (b) PCR-based restriction fragment length polymorphism (PCR-RFLP) of rDNA loci, (c) random amplification of polymorphic DNA (RAPD), (d) denaturing gradient gel electrophoresis (DGGE), and (e) pulsed field gel electrophoresis (PFGE) (Cafarchia et al., 2011). Molecular methods suitable for identification of *Malassezia* species in clinical mycological laboratories are described in Table 6.3.

Matrix-assisted laser desorption and ionization time-of-flight mass spectrometry (MALDI-TOF-MS) has been used for identification of various microorganisms and is reported to be a rapid and reliable diagnostic tool to identify environmental and clinically important yeasts. However, for *Malassezia* species there are only four main spectra projections present in the Biotyper library (MALDI Biotyper 3.1, Bruker Daltonics, Bremen, Germany), one being *Malassezia furfur* and three being *Malassezia pachydermatis*, which are stored in the CBS-KNAW Collection and DSMZ Collection (Bruker Daltonics). However, some studies have indicated that mass spectrometry is promising to identify *Malassezia* species through the construction and validation of in-house databases (Kolecka et al., 2014; Yamamoto et al., 2014). MALDI-TOF MS reference spectra obtained from type and standard strains through subculture of fresh colonies on CHROMagar *Malassezia* or modified Leeming and Notman culture medium should be added to the existing database. The protocol for discriminating the genus *Malassezia* and subsequent identification of clinical isolates is as follows:

1. Add 300 μL of water to each Eppendorf microfuge tube.
2. Transfer a large, fresh (48–72 h) colony of *Malassezia* to the tube (more than one colony may need to be transferred if yeast growth is small; choose isolated colonies) and vortex thoroughly.
3. Add 900 μL of ethanol and vortex thoroughly.
4. Centrifuge at maximum speed (16,000 *g*) for 2 minutes.

5. Remove all ethanol with pipette (tubes may be placed in a vacuum concentrator or left at room temperature to complete the evaporation process if necessary).
6. Resuspend the cell pellet in 50 μL of 70% formic acid (if only a small amount of yeast is available, decrease the formic acid volume to 10 μL) and vortex thoroughly.
7. Add 50 μL of 100% acetonitrile (if only a small amount of yeast is available, decrease the acetonitrile volume to 10 μL) and vortex thoroughly.

NOTE: The volumes of 70% formic acid and acetonitrile must be equal.

8. Centrifuge at maximum speed (16,000 g) for 2 minutes.
9. Pipette 1 μL of supernatant onto a polished steel target plate and allow it to air dry.
10. Overlay with 1 μL of matrix solution (&-cyano-4-hydroxycinnamic acid) to cover each sample spot; air dry (ensure target is completely dry before it is analyzed by MALDI-TOF-MS).

6.5 Conclusions

Clinical management of superficial *Malassezia* infections does not require normal mycological diagnosis, and *Malassezia* cultivation is not used routinely. However, in recent years, many reports have described *Malassezia* as agents of severe deep infections. Therefore, mycological laboratories should be prepared to respond to this special demand from clinicians. For this purpose, molecular tests or MALDI-TOF mass spectrometry can be used.

References

Akaza, N., Akamatsu, H., Takeoka, S. et al. 2012. *Malassezia globosa* tends to grow actively in summer conditions more than other cutaneous *Malassezia* species. *J Dermatol* 39: 613–661.

Arendrup, M. C., Boekhout, T., Akova, M. et al. 2014. ESCMID and ECMM joint clinical guidelines for the diagnosis and management of rare invasive yeast infections. *Clin Microbiol Infect* 20: 76–98.

Ashbee, H. R., Evans, E. G. V. 2002. Immunology of diseases associated with *Malassezia* species. *Clin Microbiol Rev* 15: 21–57.

Bailon, H. 1889. *Traité de Botanique Médicale Cryptogamique*. Paris: Octave Doin, pp. 234–235.

Bruker Daltonics. Extended spectra database for microorganism identification by MALDI. http://spectra.folkhalsomyndigheten.se/spectra/database/bruker.action. Bremen, Germany. Accessed December 22, 2017.

Cabañes, F. J., Coutinho, S. D., Puig, L., Bragulat, M. R., Castellá, G. 2016. New lipid-dependent *Malassezia* species from parrots. *Rev Iberoam Micol* 33: 92–99.

Cabañes, F. J., Hernández, J. J., Castellá, G. 2005. Molecular analysis of *Malassezia sympodialis*-related strains from domestic animals. *J Clin Microbiol* 43: 277–283.

Cabañes, F. J., Theelen, B., Castellá, G., Boekhout, T. 2007. Two new lipid-dependent *Malassezia* species from domestic animals. *FEMS Yeast Res* 7: 1064–1076.

Cabañes, F. J., Vega, S., Castellá, G. 2011. *Malassezia cuniculi* sp. nov., a novel yeast species isolated from rabbit skin. *Med Mycol* 49: 40–48.

Cafarchia, C., Gasser, R. B., Figueredo, L. A., Latrofa, M. S., Otranto, D. 2011. Advances in the identification of *Malassezia*. *Mol Cell Probes* 25: 1–7.

Cafarchia, C., Gasser, R. B., Latrofa, M. S., Parisi, A., Campbell, B. E., Otranto, D. 2008. Genetic variants of *Malassezia pachydermatis* from canine skin: Body distribution and phospholipase activity. *FEMS Yeast Res* 8: 451–459.

Cafarchia, C., Latrofa, M. S., Testinia, G. et al. 2007. Molecular characterization of *Malassezia* isolates from dogs using three distinct genetic markers in nuclear DNA. *Mol Cell Probes* 21: 229–238.

Cornu, M., Goudjil, S., Kongolo, G. et al. 2018. Evaluation of the (1,3)-β-D-glucan assay for the diagnosis of neonatal invasive yeast infections. *Med Mycol* 56: 78–87.

el-Hefnawi, H., el-Gothamy, Z., Refai, M. 1971. Studies on pityriasis versicolor in Egypt. I. Incidence. *Mykosen* 14: 225–231.

Findley, K., Oh, J., Yang, J. et al. 2013. Human skin fungal diversity. *Nature* 498: 367–370.

Gaitanis, G., Magiatis, P., Hantschke, M., Bassukas, I. D., Velegraki, A. 2012. The *Malassezia* genus in skin and systemic diseases. *Clin Microbiol Rev* 25: 106–141.

Gaitanis, G., Magiatis, P., Stathopoulou, K., Bassukas, I. D., Alexopoulos, E. C., Velegraki, A., Skaltsounis, A. L. 2008. AhR ligands, malassezin, and indolo[3,2-b]carbazole are selectively produced by *Malassezia furfur* strains isolated from seborrheic dermatitis. *J Invest Dermatol* 128: 1620–1625.

Gaitanis, G., Robert, V., Velegraki, A. 2006. Verifiable single nucleotide polymorphisms of the internal transcribed spacer 2 region for the identification of 11 *Malassezia* species. *J Dermatol Sci* 43: 214–217.

Glatz, M., Bosshard, P. P., Hoetzenecker, W., Schmid-Grendelmeier, P. 2015. The role of *Malassezia* spp. in atopic dermatitis. *J Clin Med* 4: 1217–1228.

Guillot, J., Deville, M., Berthelemy, M., Provost, F., Guého, E. 2000. A single PCR-restriction endonuclease analysis for rapid identification of *Malassezia* species. *Lett Appl Microbiol* 31: 400–403.

Guillot, J., Guého, E., Chévrier, G., Chermette, R. 1997. Epidemiological analysis of *Malassezia pachydermatis* isolates by partial sequencing of the large subunit ribosomal RNA. *Res Vet Sci* 62: 22–25.

Guillot, J., Lesourd, E. G. M., Midgley, G., Chévrier, G., Dupont, B. 1996. Identification of *Malassezia* species: A practical approach. *J Mycol Méd* 6: 103–110.

Guého, E., Midgley, G., Guillot, J. 1996. The genus *Malassezia* with description of four new species. *Antonie van Leeuwenhoek.* 69: 337–355.

Gupta, A. K., Boekhout, T., Theelen, B., Summerbell, R., Batra, R. 2004. Identification and typing of *Malassezia* species by amplified fragment length polymorphism and sequence analyses of the internal transcribed spacer and large-subunit regions of ribosomal DNA. *J Clin Microbiol* 42: 4253–4260.

Harada, K., Saito, M., Sugita, T., Tsuboi, R. 2015. *Malassezia* species and their associated skin diseases. *J Dermatol* 42: 250–257.

Hirai, A., Kano, R., Makimura, K. et al. 2004. *Malassezia nana* sp. nov, a novel lipid-dependent yeast species isolated from animals. *Int J Syst Evol Microbiol* 54: 623–627.

Hurwitz, S. 1981. Skin disorders due to fungi, pp. 287–288. In S. Hurwitz (ed.), *Clinical Pediatric Dermatology.* Philadelphia: The W. B. Saunders Co.

Kaneko, T., Murotani, M., Ohkusu, K., Sugita, T., Makimura, K. 2012. Genetic and biological features of catheter-associated *Malassezia furfur* from hospitalized adults. *Med Mycol* 50: 74–80.

Kolecka, A., Khayhan, K., Arabatzis, M. et al. 2014. Efficient identification of *Malassezia* yeasts by matrix-assisted laser desorption ionization-time of flight mass spectrometry (MALDI-TOF MS). *Br J Dermatol* 170: 332–341.

Lacaz, C. S., Porto, E., Martins, J. E. C., Vaccari, E. V., Melo, N. T. 2002. *Tratado de Micologia Médica.* 9 ed. São Paulo: Sarvier.

Makimura, K., Tamura, Y., Kudo, M., Uchida, K., Saito, H., Yamaguchi, H. 2000. Species identification and strain typing of *Malassezia* species stock strains and clinical isolates based on the DNA sequences of nuclear ribosomal internal transcribed spacer 1 regions. *J Med Microbiol* 49: 29–35.

Marcon, M. J., Powell, D.A. 1992. Human infections due to *Malassezia* spp. *Clin Microbiol Rev* 5: 101–119.

Marples, M. 1965. *The Ecology of the Human Skin.* Springfield, IL: Bannerstone House.

Mayser, P., Haze, P., Papavassilis, C., Pickel, M., Gruender, K., Guého, E. 1997. Differentiation of *Malassezia* species: Selectivity of cremophor EL, castor oil and ricinoleic acid for *M. furfur*. *Br J Dermatol* 137: 208–213.

Oh, B. H., Song, Y. C., Lee, Y. W., Choe, Y. B., Ahn, K. J. 2009. Comparison of nested PCR and RFLP for identification and classification of *Malassezia* yeasts from healthy human skin. *Ann Dermatol* 21: 352–357.

Peleg, A. Y., Hogan, D. A., Mylonakis, E. 2010. Medically important bacterial-fungal interactions. *Nat Rev Microbiol* 8: 340–349.

Prohić, A. 2012. Psoriasis and *Malassezia* yeasts. In J. O'Daly (ed.), *Psoriasis—A Systemic Disease*, ISBN: 978-953-51-0281-6, InTechOpen, Available from: https://www.intechopen.com/books/psoriasis-a-systemic-disease/psoriasis-and-malassezia-yeasts. DOI: 10.5772/27319.

Prohić, A., Sadikovic, T. J., Krupalija-Fazlic, M., Kuskunovic-Vlahovljak, S. 2016. *Malassezia* species in healthy skin and in dermatological conditions. *Int J Dermatol* 55: 494–504.

Schmidt, A. 1997. *Malassezia furfur*: A fungus belonging to the physiological skin flora and its relevance in skin disorders. *Cutis.* 59: 21–24.

Simmons, R. B., Guého, E. 1990. A new species of *Malassezia*. *Mycol Res* 94: 1146–1149.

Sugita, T., Tajima, M., Takashima, M. et al. 2004. A new yeast, *Malassezia yamatoensis*, isolated from a patient with seborrheic dermatitis, and its distribution in patients and healthy subjects. *Microbiol Immunol* 48: 579–583.

Sugita, T., Takashima, M., Kodama, M., Tsuboi, R., Nishikawa A. 2003. Description of a new yeast species, *Malassezia japonica*, and its detection in patients with atopic dermatitis and healthy subjects. *J Clin Microbiol* 41: 4695–4699.

Sugita, T., Takashima, M., Shinoda, T. et al. 2002. New yeast species, *Malassezia dermatis*, isolated from patients with atopic dermatitis. *J Clin Microbiol* 40: 1363–1367.

Talaee, R., Katiraee, F., Ghaderi, M., Erami, M., Alavi, A. K., Nazeri, M. 2014. Molecular identification and prevalence of *Malassezia* species in pityriasis versicolor patients from Kashan, Iran. *Jundishapur J Microbiol* 7: e11561.

Veasey, J. V., Avila, R. B., Miguel, B. A. F., Muramatu, L. H. 2017. White piedra, black piedra, tinea versicolor, and tinea nigra: Contribution to the diagnosis of superficial mycosis. *An Bras Dermatol* 92: 413–416.

Velegraki, A., Cafarchia, C., Gaitanis, G., Latta, R., Boekhout, T. 2015. *Malassezia* infections in humans and animals: Pathophysiology, detection, and treatment. *PLoS Pathogens* 1: e1004523.

Weidman, F. D. 1925. Exfoliative dermatitis in the Indian rhinoceros (*Rhinoceros unicornis*), with description of a new species: *Pityrosporum pachydermatis*. In Fox H. (ed.), *Rep Lab Museum Comp Pathol Zoo Soc*, Philadelphia, pp. 36–43.

Yamamoto, M., Umeda, Y., Yo, A., Yamaura, M., Makimura, K. 2014. Utilization of matrix-assisted laser desorption and ionization time-of-flight mass spectrometry for identification of infantile seborrheic dermatitis-causing *Malassezia* and incidence of culture-based cutaneous *Malassezia* microbiota of 1-month-old infants. *J Dermatol* 41: 117–123.

7
Rhodotorula spp.

Rejane Pereira Neves, Ana Maria Rabelo de Carvalho,
Carolina Maria da Silva, Danielle Patrícia Cerqueira
Macêdo, and Reginaldo Gonçalves de Lima-Neto

Contents

7.1 Introduction

The genus *Rhodotorula* is represented by saprophytic yeasts belonging to the phylum Basidiomycota and can be recovered from many environmental sources, such as air, soil, seawater, and plants. This genus is composed of pigmented yeasts classified in the family Cryptococcaceae and includes 38 species. *Rhodotorula* species have some morphological and physiological similarities with *Cryptococcus* spp., but differ from them by the production of carotenoid pigments (giving a yellowish to red color to the colonies) and by the inability to assimilate inositol (Granero et al., 2017; Wirth & Goldani, 2012).

7.2 Predisposing factors and clinical manifestations

Previously, the genus *Rhodotorula* was considered nonpathogenic, but various species have emerged recently as opportunistic fungal pathogens that can be responsible for localized and invasive infections, especially in immunosuppressed patients (Nunes et al., 2013).

Among infections due to *Rhodotorula* spp., fungemia is the most common manifestation. In a systemic review of 128 cases of *Rhodotorula* infections, 79% were fungemia, 7% ocular infections, and 5% peritonitis, associated with continuous ambulatory peritoneal dialysis. The species that are most often related to infections in humans are *Rhodotorula mucilaginosa*, *R. glutinis*, and *R. minuta* (Pfaller et al., 2009).

Many predisposing factors have been associated with the increase of *Rhodotorula* infections, such as administration of corticosteroids and cytotoxic drugs in solid-tumor and hematological patients, abdominal surgery, use of broad-spectrum antibiotics, and invasive medical devices. Most cases have been associated with the use of central venous catheters (CVCs). However, some of the localized *Rhodotorula* infections, like ocular, peritoneal, and prosthetic joint infections, are not necessarily associated with the use of CVCs or immunosuppression (Rajmane et al., 2016; Wirth & Goldani, 2012).

Although *Rhodotorula* species are environmental yeasts, they can also colonize human epithelium as well as the respiratory and gastrointestinal tracts. However, they are generally not considered pathogenic in immunocompetent individuals, and the isolation of *Rhodotorula* from non-sterile human sites like mucous membranes has often been of questionable clinical significance, so there is a need for association between clinical and laboratorial characteristics. On the other hand, some authors have reported confirmed cases of localized infections without fungemia in both immunocompetent and immunosuppressed hosts, including ocular infection, onychomycosis, meningitis, and prosthetic joint infections (Kim, Hyun, & Ryu, 2013; Maurya et al., 2015; Wirth & Goldani, 2012).

Ocular *Rhodotorula* infections have been reported in the form of keratitis (the most common manifestation), scleritis, dacryocystitis, and endophthalmitis. In the majority of the cases, the affected individuals were generally healthy, but most of them had associated ocular trauma or surgery (Giovannini et al., 2014; Muralidhar & Sulthana, 1995; Pradhan & Jacob, 2012; Rajmane et al., 2011).

Rhodotorula spp. are also uncommon among the agents of onychomycosis, and few cases have been reported in the literature. Since *Rhodotorula* spp. are normally found in soil, it is believed that nail infection cases are associated with people who work with soil and plants (Maurya et al., 2015; Uludag et al., 2014).

Additionally, species of *Rhodotorula* are rarely associated with meningitis. All cases reported until now have occurred in patients with HIV infection or hematological malignancies, and most of them were considered to be healthcare associated. Usually the mycosis is accompanied by fever, with a variable outcome. In some reported cases, the disease was only discovered postmortem (Loss et al., 2011; Nor et al., 2015; Thakur et al., 2007; Tsiodras et al., 2014).

Instead, *Rhodotorula* fungemia is the most common manifestation, and has been associated with high morbidity and mortality rates (up to 20%). Usually it has been reported in patients with CVCs, damage to the normal anatomic barriers, and immunosuppression. Due to the ability of *Rhodotorula* species to adhere to plastic surfaces, vascular access devices provide the necessary surfaces for biofilm formation and are responsible for the majority of bloodstream infections in humans (Miglietta et al., 2015). Although *Rhodotorula* fungemia occurs mainly in immunocompromised hosts, Kim et al. (2013) reported a case of *R. mucilaginosa* hematological infection in an immunocompetent patient with prolonged use of a CVC.

7.3 Laboratorial diagnosis

Previously regarded as non-pathogenic, *Rhodotorula* species have recently emerged as opportunistic pathogens with the ability to colonize and infect susceptible patients. Most cases of *Rhodotorula* infection are fungemia associated with catheters, endocarditis, and meningitis. Non-systemic *Rhodotorula* infections such as endophthalmitis and peritonitis have been reported in immunocompetent patients (Lunardi et al., 2006).

Isolation of *Rhodotorula* species from non-sterile sites such as skin, sputum, or stool is more likely due to colonization or contamination, and treatment in such cases should only be started in the presence of symptoms strongly suggestive of infection and after other causes have been excluded (Hagan et al., 1995).

The recovery of *Rhodotorula* species from a sterile site, such as blood, peritoneal fluid, or cerebrospinal fluid, is usually indicative of infection. The clinician should maintain a high degree of suspicion in such cases, especially if the patient has no suggestive symptoms of infection. Morphological and biochemical confirmation of the diagnosis should be sought, since yeast cells can usually be seen in microscopic examination (Larone, 2002).

In cases of meningitis, the cerebrospinal fluid typically shows lymphocytic pleocytosis, and India ink stain can reveal encapsulated budding yeast cells (Baradkar & Kumar, 2008). It may be difficult to morphologically differentiate *Rhodotorula* meningitis from cryptococcal meningitis, and a case of false-positive latex agglutination test (LAT) for cryptococcal antigen has been reported. Characteristic pigmentation of the colonies, confirmatory biochemical test results (e.g., absence of carbohydrate fermentation, production of urease), and absence of ballistospore formation are indicators for specific microbial identification.

It should be noted that that isolates of *Rhodotorula* have been found to cross-react with the *Candida glabrata/Candida krusei* probe in the commercially available fluorescence *in situ* hybridization (FISH) test for species identification in positive blood cultures (Hall et al., 2012).

Panfungal PCR allows highly sensitive and specific detection and identification of a wide spectrum of fungal pathogens in blood samples, including *Rhodotorula*. However, it is difficult to predict when these techniques will be incorporated in conventional clinical practice (Kaur et al., 2007).

Since a central venous catheter is often involved, the catheter tip should be cultured when removed so as to capture cases where blood cultures are falsely negative. *Rhodotorula* isolates are easily recognizable in the laboratory due to their distinctive orange to salmon-colored colonies, morphology, formation of rudimentary hyphae, and urease production (Wirth & Goldani, 2012).

Antigen detection has not been reported in clinical cases of systemic *Rhodotorula* infections. β1-3-D-glucan has been detected in the supernatant from isolates of *R. mucilaginosa* at an average concentration of two-thirds of that of *Candida* species. Whether the β1-3-D-glucan test would be useful as a surrogate marker for invasive *Rhodotorula* infection remains to be investigated (Odabasi et al., 2006).

7.3.1 Mycological diagnosis

The collection of clinical specimens should be performed according to the location of the lesion. There should always be more caution in the interpretation of the results when it comes to non-sterile body sites. Since *Rhodotorula* species are emerging yeasts that are isolated from several environmental sources, it is essential for the clinical samples to be collected aseptically and for the culture to be seeded in duplicate (Miceli et al., 2011).

Samples should be collected after cleansing the area with 70% isopropyl alcohol to prevent contamination. For the diagnosis of onychomycosis, nail scales should be collected. Once the specimen has been obtained, office microscopy can be performed by preparing the samples with 10%–20% potassium hydroxide (KOH) solution. The KOH will dissolve keratin, leaving the fungal cell intact (Westerberg & Voyack, 2013).

The culture should be grown on Sabouraud agar medium, maintained at a temperature between 25°C–28°C, and observed for up to 7 days. Two separate collections should be carried out on different days (Lacaz et al., 2002).

In case of possible fungemia, blood samples should be collected aseptically and seeded in duplicate, one being maintained at 25°C–28°C and the other at 37°C for 7 days. The need to maintain cultures at different temperatures is associated with the physiology of the fungus. To cause parasitism, the fungus is adapted to grow at body temperature, so it must be able to multiply at 37°C (Lacaz et al., 2002).

7.3.1.1 Microscopic and culture findings – In cases of infection by *Rhodotorula*, yeast cells are checked by direct examination. The colonies on Sabouraud agar are moist, glistering, smooth to mucoid, and salmon pink to coral red in color (Figure 7.1A). Microscopically, yeast cells are short and ovoid, with or without pseudohyphae (Figure 7.1B). The macroscopic and microscopic characteristics of *Rhodotorula* species involved in human infections are described in Table 7.1.

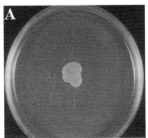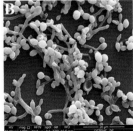

Figure 7.1   (A) *Rhodotorula* sp. colony that is moist, glistering, mucoid, and salmon pink after 5 days of growth on Sabouraud agar at 25°C. (B) Scanning electron micrograph of *Rhodotorula* sp. showing budding and isolated globose yeast cells, with pseudohyphae.

Table 7.1   Macroscopic and Microscopic Characteristics of *Rhodotorula* Species Involved in Human Infections

Rhodotorula Species	Macroscopic Characteristics of the Culture	Microscopic Characteristics of the Culture
R. glutinis	Colonies on Sabouraud dextrose agar at 25°C are pink and smooth and have a moist appearance.	On cornmeal following 72 hours of incubation at 25°C, only globose yeast cells are produced (3–5 μm in diameter). Pseudohyphae are occasionally seen.
R. minuta	Colonies on Sabouraud dextrose agar at 25°C are pink, smooth, and glossy but not mucoid.	On cornmeal following 72 hours of incubation at 25°C, only globose yeast cells are produced (2.3–6.5 μm in diameter).
R. mucilaginosa	Colonies on Sabouraud dextrose agar at 25°C range from orange to pink to coral and are smooth to rugoid and dull to glistening.	On cornmeal following 72 hours of incubation at 25°C, only globose yeast cells are produced (2.5–10 μm in diameter).

Source: Kurtzman, C. P., Fell, J. W. 2000. *The Yeasts. A Taxonomic Study*. Amsterdam, The Netherlands: Elsevier Scientific B.V.

7.3.2 Identification tools

Conventional phenotypic methods are limited in accuracy and consistency for species identification of emergent pathogens like *Rhodotorula* species. So, methods that guarantee appropriate identification are necessary for proper treatment (Nunes et al., 2013).

Fungal identification in routine laboratories is performed using automated systems (e.g., Vitek 2). Automated methods are highly accurate for the identification of common human yeast pathogens such as *Candida* species. However, they are unable to distinguish genetically similar species and have limited accuracy for identification of emergent pathogens, such as *Rhodotorula* sp. (Chagas-Neto et al., 2008; De Almeida et al., 2008).

In this context, Nunes et al. (2013) confirmed that internal transcribed spacer (ITS 1 (5′-TCCGTAGGTGAACCTGCGG-3′)/(ITS4 5′-TCCGTAGGTGAACCTGCGG-3′)) sequencing is a reliable method for the identification of *Rhodotorula* species, and is more accurate than conventional phenotypic tests. Table 7.2 shows the target regions and primers that are used for amplification and sequencing of *Rhodotorula* genes.

Furthermore, matrix-assisted laser desorption and ionization time-of-flight mass spectrometry (MALDI-TOF-MS) has been used for identification of various microorganisms and is reported to be a rapid and reliable diagnostic tool to identify environmental and clinically important yeasts, including *Rhodotorula* species. There are 17 main spectra projections present in the Biotyper library (MALDI Biotyper 3.1, Bruker Daltonics, Bremen, Germany): one *Rhodotorula acheniorum*, one *Rhodotorula bacarum*, one *Rhodotorula bogoriensis*, two *Rhodotorula glutinis*, three *Rhodotorula minuta*, eight *Rhodotorula mucilaginosa*, and one *Rhodotorula pustule*, which identify *Rhodotorula* isolates obtained from blood, soil, water, and air samples (Bruker Daltonics, 2017). The MALDI-TOF MS spectra should be obtained from analysis of three fresh colonies

Table 7.2   Target Regions and Primers Used for Amplification and Sequencing of *Rhodotorula* Genes

Target Region	Primers	Reference
ITS	ITS 1 (5′-TCCGTAGGTGAACCTGCGG-3′) ITS4 (5′-TCCGTAGGTGAACCTGCGG-3′)	Nunes et al. (2013)
D1/D2	NL-1 (5′-GCATATCAATAAGCGGAGGAAAAG-3′) NL-4 (5′-GGTCCGTGTTTCAAGACGG-3′)	Jamali et al. (2016)

Abbreviations: ITS, Internal transcribed spacer region of ribosomal DNA (rDNA); D1/D2, variable domains of the large rDNA subunit (28S).

subcultured on yeast extract peptone dextrose medium. The protocol for identification of the main *Rhodotorula* species follows:

1. Add 300 μL of water to each Eppendorf microfuge tube.
2. Transfer a large, fresh (24–48 h) colony of *Rhodotorula* to the tube (more than one colony may need to be transferred if the colony is small; choose isolated colonies) and vortex thoroughly.
3. Add 900 μL of ethanol and vortex thoroughly.
4. Centrifuge at maximum speed (16,000 *g*) for 2 minutes.
5. Remove all ethanol with a pipette (tubes may be placed in a vac concentrator or left at room temperature to complete the evaporation process if necessary).
6. Resuspend the cell pellet in 50 μL of 70% formic acid (if only a small amount of yeast is available, decrease the formic acid volume to 10 μL) and vortex thoroughly.
7. Add 50 μL of 100% acetonitrile (if only a small amount of yeast is available, decrease the acetonitrile volume to 10 μL) and vortex thoroughly.

   NOTE: The volume of 70% formic acid and acetonitrile must be equal.

8. Centrifuge at maximum speed (16,000 *g*) for 2 minutes.
9. Pipette 1 μL of supernatant onto a polished steel target plate and allow to air dry.
10. Overlay 1 μL of matrix solution (&-cyano-4-hydroxycinnamic acid) to cover each sample spot; air dry (ensure target is completely dry before it is inserted for MALDI).

References

Baradkar, V. P., Kumar, S. 2008. Meningitis caused by *Rhodotorula mucilaginosa* in human immunodeficiency virus seropositive patient. *Ann Indian Acad Neurol* 11:245–247.

Bruker Daltonics. Extended spectra database for microorganism identification by MALDI. http://spectra.folkhalsomyndigheten.se/spectra/database/bruker.action. Bremen, Germany. Accessed December 22, 2017.

Chagas-Neto, T. C., Chaves, G. M., Colombo, A. L. 2008. Update on the genus *Trichosporon*. *Mycopathologia* 166: 121–132.

De Almeida, G. M. D., Costa, S. F., Melhem, M. et al. 2008. *Rhodotorula* spp. isolated from blood cultures: Clinical and microbiological aspects. *Med. Mycol.* 46: 547–556.

Giovannini, J., Lee, R., Zhang, S. X., Jun, A. S., Bower, K. S. 2014. *Rhodotorula* keratitis: A rarely encountered ocular pathogen. *Case Rep Ophthalmol* 5: 302–310.

Granero, V., Peyronel, E., Mensa, N., Liuzzi, N., Costa, C., Cavallo, M. R. 2017. An unusual case of *Rhodotorula mucilaginosa* fungaemia in a cancer patient. *Microbiologia Medica* 32: 6827.

Hagan, M. E., Klotz, S. A., Bartholomew, W., Potter, L., Nelson, M. A. 1995. Pseudoepidemic of *Rhodotorula rubra*: A marker for microbial contamination of the bronchoscope. *Infect Control Hosp Epidemiol* 16: 727–728.

Hall, L., Le Febre, K. M., Deml, S. M., Wohlfiel, S. L., Wengenack, N. L. 2012. Evaluation of the yeast traffic light PNA FISH probes for identification of *Candida* species from positive blood cultures. *J Clin Microbiol* 50: 1446–1448.

Jamali, S., Gharaei, M., Abbasi, S. 2016. Identification of yeast species from uncultivated soils by sequence analysis of the hypervariable D1/D2 domain of LSU–rDNA gene in Kermanshah province, Iran. *Mycol Iran* 3: 87–98.

Kaur, R., Wadhwa, A., Agarwal, S. K. 2007. *Rhodotorula mucilaginosa*: An unusual cause of oral ulcers in AIDS patients. *AIDS* 21: 1068–1069.

Kim, H. A., Hyun, M., Ryu, S. 2013. Catheter-associated *Rhodotorula mucilaginosa* fungemia in an immunocompetent host. *Infect Chemother* 45: 339–342.

Kurtzman, C. P., Fell, J. W. 2000. *The Yeasts. A Taxonomic Study.* Amsterdam, The Netherlands: Elsevier Scientific B.V.

Lacaz, C. S., Porto, E., Martins, J. E. C., Vaccari, E. V., Melo, N. T. 2002. *Tratado de Micologia Médica.* 9th ed. São Paulo: Sarvier.

Larone, D. H. 2002. *Medically Important Fungi.* 4th ed. Washington, DC: ASM Press.

Loss, S. H., Antonio, A. C., Roehrig, C., Castro, P. S., Maccari, J. G. 2011. Meningitis and infective endocarditis caused by *Rhodotorula mucilaginosa* in an immunocompetent patient. *Rev Bras Ter Intensiva* 23: 507–509.

Lunardi, L. W., Aquino, V. R., Zimerman, R. A., Goldani, L. Z. 2006. Epidemiology and outcome of *Rhodotorula* fungemia in a tertiary care hospital. *Clin Infect Dis* 43: 60–63.

Maurya, V., Khatri, P. K., Meena, S. et al. 2015. Onychomycosis caused by *Rhodotorula glutinis*: A case report. *Int J Curr Microbiol App Sci* 4: 331–335.

Miceli, M. H., Díaz, J. A., Lee, S. A. 2011. Emerging opportunistic yeast infections. *Lancet Infect Dis* 11: 142–151.

Miglietta, F., Faneschi, M. L., Braione, A. et al. 2015. Central venous catheter-related fungemia caused by *Rhodotorula glutinis*. *Med Mycol J* 56: E17–E19.

Muralidhar, S., Sulthana, C. M. 1995. *Rhodotorula* causing chronic dacryocystitis: A case report. *Indian J Ophthalmol* 43: 196–198.

Nor, F. M., Tan, L. H., Na, S. L., Ng, K. P. 2015. Meningitis caused by *Rhodotorula mucilaginosa*. *Mycopathologia* 180: 95–98.

Nunes, J. M., Bizerra, F. C., Ferreira, R. C., Colombo, A. L. 2013. Molecular identification, antifungal susceptibility profile, and biofilm formation of clinical and environmental *Rhodotorula* species isolates. *Antimicrob Agents Chemother* 57: 382–389.

Odabasi, Z., Paetznick, V. L., Rodriguez, J. R., Chen, E., McGinnis, M. R., Ostrosky-Zeichner, L. 2006. Differences in b-glucan levels in culture supernatants of a variety of fungi. *Med Mycol* 44: 267–272.

Pfaller, M. A., Dickerna, D. J., Merz, W. G. 2009. Infections caused by non-*Candida*, non-*Cryptococcus* yeasts. In: Anaissie, E. J., McGinnis, M. R., Pfaller, M. A. (eds). *Clinical Mycology*. 2nd ed. London: Churchill Livingstone Elsevier; 258–259.

Pradhan, Z. S., Jacob, P. 2012. Management of *Rhodotorula* scleritis. *Eye (Lond)* 26: 1587.

Rajmane, V. S., Rajmane, S. T., Ghatole, M. P. 2011. *Rhodotorula* species infection in traumatic keratitis—A case report. *Diagn Microbiol Infect Dis* 71: 428–429.

Rajmane, V. S., Rajmane, S. T., Kshirsagar, A. Y., Patil, V. C. 2016. *Rhodotorula mucilaginosa* bloodstream infection in a case of duodenal perforation. *Avicenna J Clin Microbiol Infect* 3: e35434.

Thakur, K., Singh, G., Agarwal, S., Rani, L. 2007. Meningitis caused by *Rhodotorula rubra* in a human immunodeficiency virus infected patient. *Indian J Med Microbiol* 25: 166–168.

Tsiodras, S., Papageorgiou, S., Meletiadis, J. et al. 2014. *Rhodotorula mucilaginosa* associated meningitis: A subacute entity with high mortality. Case report and review. *Med Mycol Case Rep* 6: 46–50.

Uludag, A. H., Meral, T., Aribas, E. T. et al. A case of onychomycosis caused by Rhodotorula glutinis. Case reports in dermatological medicine. *Hindawi* 2014. Doi: 10.1155/2014/563261.

Westerberg, D. P., Voyack, M. J. 2013. Onychomycosis: Current trends in diagnosis and treatment. *Am Fam Physician* 88: 762–770.

Wirth, F., Goldani, L. Z. 2012. Epidemiology of *Rhodotorula*: An emerging pathogen. *Interdiscip Perspect Infect Dis* 2012: 465717.

8
Dermatophytes

Germana Costa Paixão, Marcos Fábio Gadelha Rocha,
Débora de Souza Colares Maia Castelo-Branco, Raimunda
Sâmia Nogueira Brilhante, and José Júlio Costa Sidrim

Contents

8.1 Historical aspects

Dermatophytes belong to the oldest groups of microorganisms that were recognized as agents of human diseases. Medical mycology started with the study of these fungi by Robert Remak in 1839, when he elucidated the etiology of "Favus." In 1842, David Gruby independently discovered the etiological agent of "Favus," created the genus *Microsporum*, and reaffirmed the fungal etiology of all the tineas. Despite the findings of these and other researchers, it took more than a half-century for medical mycology to emerge from the general ostracism that marked the ensuing years. In 1910, Raymond Jacques Andrien Sabouraud, an eminent dermatologist trained at the Pasteur Institute, published a treatise on medical mycology, *Les Teignes*, a landmark in the field that aimed to standardize the existing knowledge at the time regarding the most often discussed topic in mycology, the dermatophytes.

Sixteen species of dermatophytes associated with human diseases were introduced between 1870 and 1920, and Sabouraud, with a magisterial vision, defined a new classification system, subdividing these species into four genera: *Achorion*, *Trichophyton*, *Microsporum*, and *Epidermophyton*. During the decades that followed, based on confusing concepts and application of new methodological standards, there was an explosion of new species to which recombined names were assigned, leading to the introduction of many taxa that are now considered synonyms of previously described species.

In 1934, C.W. Emmons, revising the criteria for classifying fungi at the time, proposed a new classification for dermatophytes based on microscopic aspects of the conidia and the ability of these fungi to grow on special media, mainly those based on cereals. Emmons abolished the genus *Achorion* proposed by Sabouraud and placed the corresponding species in the genus *Trichophyton*, leading to the classification of dermatophytes into three genera: *Trichophyton*, *Microsporum*, and *Epidermophyton*.

The 1960s marked a new phase in the knowledge of dermatophytes, with the rediscovery of the sexual reproduction of some species. Thus, Donald Griffin rediscovered the perfect form of *Microsporum gypseum*, previously described by Nannizi in 1926.

In this way, fungi of the group of dermatophytes belonging to the genus *Trichophyton* that presented sexual reproduction were called *Arthroderma* when in perfect form. This was also observed among fungi of the *Microsporum* genus, which were called *Nannizzia* when in the teleomorph or perfect form. However, in 1986, Weitzman and colleagues, based on the observation that the genera *Nannizzia* and *Arthroderma* have very similar morphological traits, resolved to join them in a single genus, *Arthroderma*, following the taxonomic law that gives preference to the oldest denomination.

In the last decades of the twentieth century, it became evident that classification based on morphology had limitations and could not be used as the only characteristic for classification or identification. In response, Weitzman introduced an additional analysis in the form of physiological parameters, according to growth in Trichophyton agar media (T1 to T7), relying on the ability of strains to assimilate a panel of essential vitamins an amino acids, associated with growth temperature, liquefaction of gelatin, etc. Thus, the taxonomy of dermatophytes combined the clinical appearance, characteristics of growth in culture media, microscopic features, and physiology.

Similar to the days of Pasteur, when the axenic culture technique revolutionized microbiology, today the application of molecular methods has revolutionized the taxonomy of dermatophytes and other pathogenic fungi (Ahmadi et al., 2015; Calderaro et al., 2014; Gnat et al., 2017; Mochizuki et al., 2017). However, although the molecular approach resolves the principal characteristics of dermatophyte evolution, there are still drawbacks to this characterization. Several fungi that are clinically well established as different species of *Trichophyton* (*Trichophyton rubrum*/*T. violaceum*, *T. equinum*/*T. tonsurans*) and *Microsporum* (*Microsporum audouinii*/*M. canis*/*M. ferrugineum*) are indistinguishable in multilocus analyses, as small ambiguities in gene sequences or lack of data conceal small differences between species (De Hoog et al., 2017).

During the decades of dual nomenclature, species could have two types, but since 2013, under the new nomenclature rules for fungi, the name anamorph or teleomorph now refers to the same sample of the original type, instead of having distinct names.

Nowadays, the perspective for the phylogenetic classification of the dermatophytes exists, and, since 2015, several authors have dedicated their efforts to better elucidating the phylogenetic tree of these fungi, based on comprehensive analysis of the ITS region of ribosomal DNA and multilocus analyses, enabling the distinction of seven monophyletic and polyphyletic clades, encompassing the genera *Arthroderma*, *Epidermophyton*, *Lophophyton*, *Nannizia*, *Paraphyton*, *Trichophyton*, and *Ctenomyces* (De Hoog et al., 2017; Bouchara et al., 2017, Gräser et al., 2018). However, the denomination dermatophyte, in the strict sense of the word, can only be used for species belonging to the genera *Microsporum*, *Trichophyton*, and *Epidermophyton*, which are keratinophilic and able to cause diseases in humans and animals, with the other species being classified as a related group, not considered true dermatophytes.

Finally, this interest in phylogenetic studies renewed the global interest in investigating the taxonomy of dermatophytes. This prompted the International Society for Human and Animal Mycology (ISHAM) to create a working group on dermatophytes in 2017, to stimulate and coordinate an international network dedicated to studying these fungi. Therefore, it is possible that in the near future new information will be added for the systematic study of these fungi.

8.2 Epidemiology

In recent years, the literature has been disclosing the increasing incidence of fungal infections, with dermatophytoses being the main cause of this increase. Several factors have been suggested for the rising incidence of fungal skin infections, among them better laboratory and clinical diagnosis, increased survival of patients with immunosuppressive diseases, and the use of drugs that exert selective pressure and permit the installation of normally saprophytic microorganisms. Therefore, dermatophytes, the traditional skin pathogens, have a greater possibility of causing infectious processes. Indeed, they have become the most often isolated fungal group in human diseases. Based on the estimated prevalence, it is believed that 10%–15% of humans will be infected with these fungi during their lifetime (Nenoff et al., 2014a).

Concerning geographic distribution, dermatophytes are cosmopolitan, but there are marked regional variations regarding the types and frequency of the isolated species. Many species have global distribution, such as *T. rubrum* and *Epidermophyton floccosum*. Others are restricted to specific geographic regions, for example, *M. ferrugineum* is found in Asia and Africa, *T. megninii* in Europe, and *T. concentricum* in South America and the Pacific islands (Coulibaly, 2018; Bouchara et al., 2017).

The international literature indicates *T rubrum* as the most commonly isolated dermatophyte species, followed by *T. violaceum* and *T. mentagrophytes* (Kupsch et al., 2016; Nenoff et al., 2014a). However, these findings cannot be taken as the absolute truth, since a good part of this literature focuses on geographic areas where the geoclimatic and social conditions are very different from other regions that have been less studied, directly influencing the spectrum of isolated dermatophyte species.

Another factor that should be considered in the epidemiology of dermatophytoses, besides geography, is the period covered by the study, since the pattern and frequency of dermatophytoses can change over the years. For example, a study conducted in France covering the period from 1956–1980 found that the dermatophytes most often isolated from scalp infections were *M. canis*, with 66.8%, followed by *T. violaceum*, with 12.1%. In another French study, covering the years 1981–1989, the most often isolated dermatophytes from scalp infections were *T. soudanense*, with 54.1%, followed by *M. langeronii*, with 26.5%. A possible explanation for this change in the pattern of isolated dermatophytes is the population flow between France and its former African colonies (Coulibaly et al., 2018).

Hence, it can be concluded that several factors, such as climate conditions, social practices, and human population mobility, influence the epidemiology of dermatophytoses; therefore, retrospective and prospective studies should be conducted to revalidate the epidemiological data in the field.

Epidermophyton floccosum, *Microsporum audouinii*, and *Trichophyton schoenleinii* have been the main pathogens causing superficial fungal diseases in the past 100 years, but their frequencies have declined dramatically since the 1950s, and are now limited to some developing countries. On the other hand, other species, like *Trichophyton tonsurans*, have become the main agent of tinea capitis in several European countries, especially the United Kingdom and France, as well as in the United States, Japan, Africa, and the Caribbean, leading to its classification as an emerging pathogen (Bouchara et al., 2017).

At present, other dermatophytes, such as *Trichophyton violaceum*, *Trichophyton verrucosum*, and *Microsporum ferrugineum*, are considered to be endemic in some parts of Africa, Asia, and Europe. However, *Trichophyton rubrum*, *Trichophyton interdigitale*, *Trichophyton tonsurans*, and *Microsporum canis* are the most cosmopolitan species, presenting the highest prevalence in the majority of the countries.

Therefore, this chapter will cover the most cosmopolitan dermatophytes, along with some endemic species.

8.3 Clinical manifestations

The clinical aspects of dermatophytic lesions are highly varied and result from the combination of the destruction of keratin with an inflammatory response, which is more or less intense depending on the parasite-host relationship. Therefore, the clinical characteristics of the lesions are strongly correlated with three factors: the species of dermatophyte involved in the infection, the affected site, and the immunological status of the host (Nenoff et al., 2014b).

There are two ways of clinically classifying dermatophytoses. One follows the English school, by which all infections are called *Tinea* (in Latin) associated with another word (also in Latin) that indicates the site of the lesion: *tinea corporis* (located on the body), *tinea capitis* (scalp), *tinea unguium* (nails), *tinea pedis* (feet), *tinea manuum* (hands), *tinea cruris* (groin), and *tinea barbae* (beard). In turn, the other clinical classification system follows the French school, classifying dermatophytoses into *Tinea* (any dermatophytic lesion on the scalp and/or beard and mustache region), epidermophytoses (dermatophytic lesions on glabrous skin), dermatophytic onychomycoses (nail lesions), and subcutaneous and deep dermatophytoses (lesions in the subcutaneous space and other deep organs, generally affecting immunocompromised patients).

T. rubrum is the main pathogen involved in infections of the skin and nails, while *M. canis*, *T. tonsurans*, and *T. violaceum* are the main species involved in *tinea capitis*. However, this pattern is not static, and population mobility, lifestyle changes, and the advent of new antifungal drugs lead to the continuous evolution of dermatophytes, emphasizing the importance of performing broad epidemiological studies to better understand the distribution dynamics of these fungi.

8.4 Laboratory diagnosis

The laboratory diagnosis of dermatophytoses, as for other fungal infections, involves three distinct phases: pre-analytical phase, consisting of indication and collection of clinical specimens; analytical phase, when the mycological analyses, per se, are performed in the laboratory and the final report is issued; and post-analytical phase, when the recovered pathogens are stored for further study (De Hoog et al., 2000; Nenoff et al., 2014b).

The correct identification of dermatophyte species should combine conventional methods based on culture growth, including morphology of colonies, characteristics of fructification (macroconidia and microconidia), and ornamentation structures and physiological and/or nutritional features, with molecular analyses by several molecular techniques, including DNA-based methods or the analysis of protein profiles by using matrix-assisted laser desorption/ionization-time-of-flight mass spectrometry (MALDI-TOF MS) (L'Ollivier and Ranque., 2017; De Hoog et al., 2017).

After collection of clinical specimens under the clinical suspicion of dermatophytosis, sample processing should start immediately. For this purpose, one or two wet-mount slides are prepared with the clarifying agents KOH (10%–40%) or K-dye. After preparing the slides, it is necessary to wait 3–5 minutes for the clarification of the clinical specimen in order to facilitate the visualization of the fungal structures. These structures, when observed, should be described considering the characteristic findings reported in the literature, to support clinical diagnosis. Hyphal characteristics should be noted regarding pigmentation, septation, and presence of arthroconidia. When the clinical specimen is a hair, the type of parasitism observed should be described, stating whether it is ectothrix or endothrix and microporic or megasporic. All these data together will support the clinical suspicion of the infection, which may later aid the mycologist in the diagnosis of the disease.

The remaining material should be seeded in three tubes containing Sabouraud agar (alone, supplemented with chloramphenicol, and supplemented with chloramphenicol and cycloheximide) and incubated for 8–15 days to detect fungal growth and maturation. However, this incubation period can be extended if there is suspicion of certain species, such as *T. concentricum* and *T. schoenleinii*, which take longer to grow. After suitable growth and the appearance of mature colonies (i.e., with well-defined patterns), a slide should be prepared with a fragment of the colony, generally from the central part, where the fungal structures are the best defined, thus facilitating the identification of fructification and/or ornamentation structures.

The optical microscopic observation involves the preparation of a wet-mount containing a fragment of the colony and one or two drops of the stain lactophenol cotton blue for the visualization of the fructification structures (macroconidia and microconidia) and ornamentation structures (racket-shaped hyphae, pectinate hyphae, spiral or tendril hyphae, favic chandeliers, etc.).

Mostly, at this point, the mycologist has enough information to issue a final diagnostic report to the clinician, who is typically anxious due to laboratory delay, although the final report will have little influence on the clinical conduct. To overcome this scenario, a good dialogue should be established between the mycologist and clinician, to "smooth out the rough edges."

Nevertheless, to the despair of the clinician and the mycologist as well, it is not always possible to identify the causative agents through microscopic examination. In such cases, the mycologist should contact the clinician to explain the diagnostic difficulties, and should continue the identification process, proceeding to the slide culture for further microscopic evaluation and biochemical tests for physiological analysis. The bottom line is that it can take up to 25 days between receiving the sample and the final identification.

8.4.1 Growth on special culture media

Any time the fungal growth in primary culture media, such as Sabouraud agar or potato agar, only leads to the observation of sterile hyphae, special culture media containing substances and ingredients that favor the development of the fructification and ornamentation structures, typical to each species, should be used. The most often used media for this purpose are rice agar and lactrimel agar.

The seeding procedure is the same used for primary media. However, instead of using the clinical specimen for the culture procedure, a small fragment of the colony obtained in the primary recovery must be used. The media are then incubated at 25°C–30°C for 5–10 days. After colony growth, a wet-mount is assembled with lactophenol cotton blue for examination under an optical microscope for the identification of the recovered fungi.

The use of rice agar is particularly useful for diagnosing fungi of the *Microsporum* genus, since this medium favors the development of typical macroconidia for each species. In turn, lactrimel agar is a culture medium that is extremely rich, leading to the appearance of numerous fructification and ornamentation structures, facilitating the identification of fungal species in general.

8.4.2 Slide culture

This technique enables detailed study of different fungal structures, including their arrangement on the hyphae. The culture slide is assembled by using a scalpel to cut squares of potato agar, measuring 5 × 5 mm, with a thickness of approximately 0.4 mm. These squares are placed on a microscope slide and are inoculated with a small fragment of the fungal colonies in the four corners of the square. Then, the block of potato dextrose agar is covered with a sterile coverslip and incubated in a sterile humid chamber for about 15–20 days at room temperature. When the growth is sufficient, the coverslip with the adhered mycelium is removed from the top of the agar block, mounted on another glass slide containing lactophenol cotton blue, and examined under an optical microscope. It is also possible to perform this observation on the original glass slide by removing the agar block from it, placing a drop of lactophenol cotton blue, and covering the material with another coverslip.

8.4.3 Urease production

The urease production test is widely used to identify dermatophytes, and is particularly useful to differentiate *Trichophyton interdigitale* from the other species of *Trichophyton*. This test is based on the ability of certain dermatophyte species to produce the enzyme urea, releasing ammonia and increasing the medium's pH, causing its color to change from yellow to dark pink.

The test involves seeding a small fragment of the dermatophyte colony on Christensen's urea agar. The tube is incubated at room temperature and interpreted after 96 hours. The result is considered positive when the medium changes from yellow to dark pink. If the medium remains yellow, the test is considered negative.

Reading and interpretation of the test should not exceed 7 or 30 days, in some rare exceptions, when fungal growth takes longer, since dermatophytes may die due to medium nutritional exhaustion, releasing alkaline metabolites that cause the color of the agar to change, leading to false positive results.

8.4.4 *In vitro* hair perforation test

This test allows distinguishing *Trichophyton interdigitale* (which has typical perforation organs) from other species of *Trichophyton*, such as *Trichophyton tonsurans* and *Trichophyton rubrum*, which do not have these organs.

There are several methods that can be used to test the ability of dermatophytes to perforate hairs. At our laboratory, the following method is used: small fragments of the colony to be analyzed are equidistantly seeded on bacteriological agar in a 70-mm-diameter Petri dish. Then, place on top of the culture fragment, within the agar, short pieces of sterilized blonde prepubertal hair, as this type of hair facilitates the visualization of fungal perforation. Incubate the agar plate at room temperature (25°C–30°C), for 7–40 days, with weekly observations to check for perforations in the hair segments. For this purpose, some hair pieces should be removed from the culture medium and placed on a slide with one or two drops of lactophenol cotton blue and covered with a coverslip. The hair pieces should be examined under an optical microscope for the presence of perforations.

8.4.5 Vitamin and nutritional tests

These tests are of great value for definitive phenotypical identification of *Trichophyton* species, since they have macromorphological and micromorphological traits that are hard to differentiate, often requiring basing the identification on differences in nutritional needs. For example, *T. verrucosum* does not grow on media containing only casein, requiring thiamine or thiamine plus inositol for growth. In contrast, *T. schoenleinii* grows indistinctly on media supplemented with vitamins or not, while *T. equinum* needs niacin and *T. tonsurans* requires thiamine.

Based on the physiological differences between dermatophyte species, all isolates belonging to the *Trichophyton* genus that cannot be identified by macromorphological or micromorphological features should be grown on culture media enriched with thiamine, histidine, niacin, and/ or inositol. These media are designated T1, T2, T3, T4, T5, T6, and T7, depending on their components.

For seeding on culture media, a fungal suspension is prepared in saline solution, with a turbidity corresponding to standard 2 on the McFarland scale, and inoculated on the agar. The agar slants are incubated at room temperature (25°C–30°C) for approximately 15 days and are evaluated by subjective and comparative quantification of the fungal growth in each tube [scale (0) for no growth up to (++) for maximum growth].

8.5 DNA-based molecular identification

The range of diagnostic tools applied to identify dermatophyte species has expanded in recent decades, especially due to the use of molecular methods, which are currently considered the gold standard. Although these methods are very precise and rapid, they are also expensive and can be complex to implement in the routine of a clinical laboratory.

The molecular identification of dermatophytes mainly relies on DNA sequencing. Although these techniques have greatly evolved, there is still a lack of consensus on result interpretation, since phenotypic identification of species does not always result in distinct molecular taxonomic entities, hampering the popularization of molecular methods among medical mycology laboratories.

The main molecular methods are divided between conventional PCR and real-time PCR, and also vary according to the DNA extraction procedures, primers, and way of analyzing the PCR products. The conventional PCR methods have the advantage of simplicity and lower cost, while real-time PCR allows the simultaneous detection of a wider spectrum of species and has lower contamination risk because it is performed in a closed system.

The first molecular studies to identify fungi used small ribosome subunits and large non-specific markers, which made precise diagnosis of species nearly impossible. To overcome these difficulties, the sequencing of more specific rDNA regions, such as region D1-D2, beta-tubulin, and ribosomal protein L10, are increasingly common and attain more precise results (De Hoog et al., 2017).

Recent advances in molecular techniques provide new ways to identify dermatophytes, including their intraspecific variations. The intraspecific subtyping and differentiation of strains make it possible to track infections, identify common sources and recurrence of infection after treatment, and analyze strain virulence and drug resistance. Analysis of mitochondrial DNA and non-transcribed ribosomal RNA genes, random amplified polymorphic DNA (RAPD), and microsatellite markers are among the main molecular methods for intraspecific subtyping and differentiation of dermatophyte strains (Mochizuki et al., 2015).

Additionally, some molecular techniques are useful to understand the phylogenetic relations between dermatophyte taxa, among them restriction fragment length polymorphism (RFLP) of mitochondrial (mt) DNA, RAPD, and sequence analysis of genes with specific functions, such as ribosomal DNA (rDNA) genes and genes that encode chitin synthase I, DNA topoisomerase II, and beta-tubulin. These techniques have led to the new classification of dermatophytes, including phylogenetic reconstructions based on the nucleotide sequence of the internal transcribed spacers (ITS). It has been shown that these classifications generally corroborate the clinical and ecological characteristics of these strains.

Therefore, progress in molecular methods has resulted in techniques for more accurate identification of dermatophytes. Nevertheless, the benefits obtained by using these molecular methods in mycology laboratories should be weighed against the relative importance of obtaining quick results and the high price of reagents and equipment.

8.6 Identification based on protein profile (matrix-assisted laser desorption/ionization-time of flight)

Matrix-assisted laser desorption/ionization-time-of-flight mass spectrometry has been increasingly applied. It was originally adapted to identify prokaryotic organisms, but new protocols have expanded its use to the analysis of some eukaryotic organisms, among them filamentous fungi and yeasts.

This method analyzes the protein profile of microorganisms and compares each obtained mass spectrum with reference mass spectra contained in databases by searching for correspondence of patterns, including peaks and intensity, by multivariate statistical analysis (Calderaro et al., 2014; L'Ollivier and Ranque, 2017).

The spectral measurements are performed within a mass variation of 3,000–8,000 Da and are recorded in positive linear mode. The raw spectra and composition of cross-correlations and autocorrelations of all the intervals, in terms of geometric means, are used as the parameters for distance between the spectra. In general, an identification index ≥ 2.0 is accepted as reliable identification at the species levels, while values ≥ 1.7 and < 2.0 are suitable for the genus level. Results < 1.7 are considered unreliable.

Even though MALDI-TOF mass spectrometry has become an increasingly powerful tool in clinical microbiology, allowing rapid identification of bacteria and yeasts, its use to identify dermatophytes remains incipient due to the lack of clear definition of species within some taxa and the lack of protein profiles for these fungi deposited in reference data banks.

Therefore, it is evident that the algorithm used to distinguish closely related dermatophyte species needs to combine phenotypic and genotypic methods.

8.7 Main species of dermatophytes

The evolution of knowledge on dermatophytoses has been marked by a large number of taxonomic classifications of dermatophytes, but approximately 10 species have been described as the most prevalent etiological agents of dermatophytoses. It is recognized that taxonomic studies in the past, solely based on phenotypical features, have led to the over-classification of fungal species. Currently, with the support of molecular biology, this is being reversed, and many species are being joined together, reducing the number of species considered pathogenic to humans.

Below, we describe the main phenotypic characteristics of the most often isolated species from human clinical specimens.

8.7.1 The *Microsporum* genus

Several *Microsporum* species are involved in human and animal infections, depending on geography and hosts' characteristics. *M. audouinii*, *M. ferrugineum*, *M. nanum*, *M. canis*, and *M. gypseum* are the most often isolated species, and *M. canis*, *M. audouinii*, *M. ferrigineum*, and *M. gypseum* (Figure 8.1A) are the most often recovered species from human infections.

8.7.1.1 *Microsporum canis* – It is a zoophilic dermatophyte transmitted to humans by several domestic animals, with young cats as its main reservoir. Clinically, it is responsible for scalp lesions that are fluorescent under a Wood's lamp, characterized by large alopecic patches, mainly in children. However, it can cause epidermophytosis, and more rarely onychomycosis.

It grows moderately quickly on Sabouraud medium, yielding mature colonies in 6–10 days. The surface of the colony is white to cream colored and has a cottony texture with discrete radial grooves and umbilical elevation. The reverse of the colony is yellowish-green, and this pigment can slightly diffuse in the medium with time, giving it a brown color. Pleomorphism often quickly occurs, with the colonies becoming thicker.

Direct microscopic observation of the colony usually reveals a large number of spindle-shaped macroconidia with thick walls and septations, varying from 5 to 7, although this can sometimes reach 15 (Figure 8.1B). Microconidia, when present, are sessile and vary in number, but they have no diagnostic value. Chlamydoconidia, nodular organs, and pectinate hyphae can also be observed at times.

The genus *Microsporum* is often suspected when observing suggestive dermatophyte colonies where only sterile hyphae are present. In this situation, the fungus in question should be immediately transferred to rice agar and observed for the appearance of fructification structures, which generally takes 6–10 days.

8.7.1.2 *Microsporum ferrugineum* – It is an anthropophilic species that is endemic to Asia (especially China and Japan), Russia, Eastern Europe, and Africa. Clinically, it is mainly involved in scalp lesions in children.

Direct examination reveals ectothrix infection, similar to *M. canis*. It grows slowly on Sabouraud agar, yielding two colony types: one has a characteristically glabrous and grooved appearance and an elevation in the center, with orange to rust color, which quickly fades when maintained in the laboratory. The other colony type is flat and white, with a leathery texture.

The micromorphology of the species is poor and it usually does not produce macroconidia or microconidia in the routine culture media. It can form characteristic bamboo hyphae.

8.7.2 The *Trichophyton* genus

This is the most often isolated genus from clinical samples. It can affect the hairless skin, scalp, and nails. Many species are imputed as being human and animal pathogens, such as: *T. rubrum, T. tonsurans, T. mentagrophytes, T. verrucosum, T. shoenleinii, T. concentricum, T. equinum, T. violaceum*, etc. With highly variable incidence and dependence on geographic conditions, these dermatophytes are responsible for nearly all cases of infections, with greater or lesser prevalence of one species or another in a given region.

The features of the *Trichophyton* spp. colonies vary according to the species, but, in general, microscopic observation reveals a large number of pyriform, oval, or round microconidia, arranged

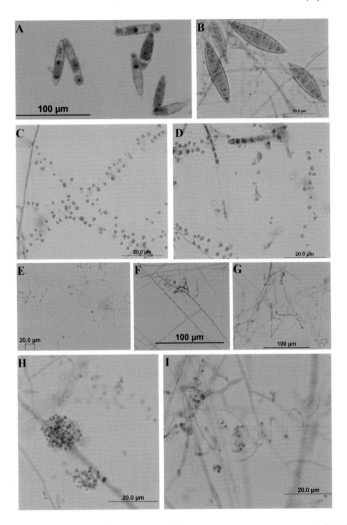

Figure 8.1 Micromorphological aspects of dermatophytes species commonly found in clinical laboratory. (A) *M. gypseum*; spiny macroconidia with thin walls and rounded ends; (B) *M. canis*; spiny macroconidia with rough thick walls and knob ends; (C) *T. rubrum*; regular pyriform microconidia arranged in acladium formation; (D) *T. rubrum*; cylindrical clavate shape macroconidia; (E–G) *T. tonsurans*; microconidia with heterogeneous aspect (small or large claviform and balloon-shaped cells); (H) *T. mentagrophytes*; round microconidia arranged in bunches; (I) *T. mentagrophytes*; spiral hyphae (ornamentation structure).

in right angles with the hyphae, bunches, or forming other structures. The macroconidia, when present, have an elongated claviform shape with a large number of septa. These fructification structures are sometimes or always absent in some species, such as *T. schöenleinii*.

The ornamentation structures are highly varied, with the presence of pectinate hyphae, spiral or tendril hyphae, racket-shaped hyphae, favic chandeliers, etc. The combination of these macromorphological and micromorphological structures is not always sufficient to identify this group of dermatophytes, a situation that poses a hurdle to the majority of those studying mycology. Therefore, it is common to encounter strains whose identification requires complementary tests, such as hair perforation, urease, and nutritional testing. These are described in the laboratory diagnosis section.

Considering the large number of species, here we only describe the most commonly isolated species (*T. rubrum*, *T. tonsurans*, *T. mentagrophytes*, and *T. verrucosum*) and two other species that cause a very distinctive clinical manifestation (*T. concentricum* and *T. shöenleinii*).

8.7.2.1 *Trichophyton rubrum* – It is a cosmopolitan anthropophilic species and is the leading cause of human dermatophytoses, according to the international literature. It is exclusively transmitted between people or by contaminated fomites. Clinically, it is associated with nearly all types of human dermatophytic infection. This is related to the fact that it is often refractory to commonly prescribed treatment, due to its strong adaptation to the human host and its greater ease of overcoming hosts' innate defenses and remaining in a residual infection, with occasional clinical recurrences.

The growth pace for this species is intermediate, taking 12–16 days to grow on Sabouraud agar.

The colonies have a cottony or velvety texture, with radial grooves, forming a small protrusion in the center, a morphology that causes frequent confusion with *T. tonsurans*. The colonies are generally white, but turn reddish with time, a color that is also found on the edges. The reverse of the colony also turns reddish with time, or different shades of brown in some strains. This pigment can generally diffuse in the medium. In some strains, this pigment is hard to discern; hence, the suspected colony should be transferred to cornmeal agar with dextrose for better observation of the color.

Optical microscopy reveals a large number of fine, regular pyriform microconidia, often arranged in a right angle with the hyphae (Figure 8.1C). Macroconidia, when present, are long with nearly cylindrical clavate type, with two to nine septations (Figure 8.1D). The presence of these macroconidia is not a definitive diagnosis of the species.

In most routine laboratories, the observation of a few structures is used for definitive diagnosis by experienced mycologists. However, sometimes identification based on phenotypic features is hard, even for experienced professionals, thus requiring complementary tests for accurate identification.

Differential diagnosis can be based on the ability of *T. rubrum* to produce pigment on cornmeal agar supplemented with 1% dextrose, the absence of perforation organs, and the variability of the urease levels. Furthermore, growth on *Trichophyton* agar nutrient media occurs with the following order of intensity: T_1, T_2, T_3, T_4, T_5 (++), T_6(+/−), and T_7. These data and those regarding other dermatophyte species are described in Table 8.1.

8.7.2.2 *Trichophyton tonsurans* – It is an anthropophilic fungus par excellence, although few cases have been reported in horses and dogs, with high prevalence in North Africa and South America. Clinically, it is responsible for scalp lesions with hair loss, but it may also cause lesions of glabrous skin and onychomycosis.

This species has intermediate growth pace, reaching maturity between 12 to 16 days of growth. The colonies, when well developed on Sabouraud agar, have a highly variable appearance. The texture can range from cottony to velvety, possibly becoming scaly with age. The relief can be apiculate, crateriform, or even cerebriform, with some radial grooves. The color of the front is highly variable, with a tendency to white or various shades of beige, while the reverse has varied shades of reddish brown, which can diffuse into the Sabouraud agar, but not in potato agar. The pigment diffusion characteristic often confounds macroscopic identification of this species with *T. rubrum*.

Optical microscopic examination reveals several microconidia arranged in a right angle with the hyphae. These microconidia have a rough aspect, as they have no homogeneity in dimension and shape, presenting themselves as small or large claviform structures, or dilated and balloon-shaped ones (Figure 8.1E through G). Intercalary chlamydoconidia, racket-shaped hyphae, and arthroconidia can also be observed in old colonies. The macroconidia, when present, are irregularly shaped and contribute little or nothing to the final species identification.

Table 8.1   Nutritional Tests to Distinguish the Most Commonly Isolated Dermatophyte Species from Humans and Animals

Species	T₁	T₂	T₃	T₄	T₅	T₆	T₇	Urease	Hair Perforation
Epidermophyton floccosum	+	+	+	+	+	+	+	F	−
Microsporum amazonicum	+	+	+	+	+	+	+	+	−
Microsporum audouini	+	+	+	+	+	+	+	V	−
Microsporum canis	++	++	++	++	++	++	++	+R	+
Microsporum gypseum	+	+	+	+	+	+	+	+	+
Microsporum nanum	+	+	+	+	++	+	+	+	+
Microsporum persicolor	+	+	+	+	+	+	+	+	+
Trichophyton concentricum	0	+	++	+	+	0		F	−
Trichophyton mentagrophytes	++	++	++	++	++	++	++	+	V
Trichophyton rubrum	++	++	++	++	++	+/−	++	V	−
Trichophyton schoenleinii	+	+	+	++	+	+	+	V	−
Trichophyton terrestre	+	+	+	+	+	+	+	+	+
Trichophyton tonsurans	+/−	+/−	+/−	+/−	+	+/−	−F	V	−
Trichophyton verrucosum	0	F	++	+	0	0	0	−	−
Trichophyton violaceum	+/−	+	+	+	+	+	+	V	−

Abbreviations: 0: no growth; +: little growth or nearly comparable with the growth control; ++: growth equal to or greater than growth control; −: negative test; +/−: little or no growth; F: very little growth; −F: negative or very little growth; +R: weak or positive reaction; V: variable results, can be positive or negative. (Adapted from de Hoog, G. S. et al. 2000. *Atlas of Clinical Fungi*. 2nd ed, Centraalbureau voor Schimmelcultures/Universitat Rovira i Virgili.)

Paradoxically, *T. tonsurans* has been found parasitizing human hair *in vivo*, but does not perforate hair *in vitro*. Other physiological aspects of *T. tonsurans* are the little or lack of urease production, as determined by incubating urease test tubes for up to 30 days; lack of growth on cornmeal agar; and little or no growth on T₁, T₂, T₃, T₄, and T₆ (+/−), but good growth on T₅ (+) (Table 8.1).

8.7.2.3 *Trichophyton interdigitale* – Until recently, the majority of authors considered *T. mentagrophytes* to have two to four varieties, with consensus being the existence of *T. mentagrophytes var. mentagrophytes* and *T. mentagrophytes var. interdigitale*. Basically, two characteristics related to ecology distinguished them: *T. mentagrophytes var. mentagrophytes* was known as a predominantly zoophilic species, while *T. mentagrophytes var. interdigitale* was an anthropophilic species par excellence.

Some significant taxonomic changes have recently been proposed for this species, and *T. interdigitale* is no longer considered a variety. Instead, it has been raised to a strictly anthropophilic species, while the zoophilic strains, formerly called *T. mentagrophytes* var. *mentagrophytes*, now constitutes the species *T. mentagrophytes*.

Besides this biological difference, these two species can present phenotypic differences, albeit not always evident, which can help differentiate them. In this context, when *T. mentagrophytes* is grown on Sabouraud agar, the colony presents a scaly or powdery texture, without remarkable elevations, sometimes forming concentric rings, with color that varies from light yellow to reddish-brown. The reverse of the colony generally has brown pigment, which can tend to wine-colored. In turn, when *T. interdigitale* is grown on Sabouraud agar, it generally forms colonies with velvety or cottony texture and light yellow color, and the reverse has light brown or red pigmentation.

Clinically, *T. interdigitale* is considered the second or third leading cause of human dermatophytoses, causing epidermophytosis, onychomycoses, scalp lesions, and interdigital plantar lesions. These lesions, when caused by *T. mentagrophytes*, often show an intense inflammatory response. The scalp lesions do not fluoresce under a Wood's lamp.

T. mentagrophytes grows rapidly, with maturation within 6–11 days after initial seeding. Microscopically, the colonies have an abundance of fructification structures. A large number of round microconidia can be observed forming bunches (Figure 8.1H). The macroconidia, when present, are generally cigar-shaped, with one to six transversal septa linked to septate hyaline hyphae. It is also common to observe a large number of ornamentation structures, such as spiral hyphae (Figure 8.1I), nodular organs, racket-shaped hyphae, and intercalary chlamydoconidia.

The physiological ability of *T. mentagrophytes* is remarkable, including intense urease production, which is observed within 3 days of growth on urea agar, while *in vitro* hair perforation test results are variable within up to 40 days. No pigment is observed when grown on cornmeal agar, and it grows poorly on T_7 (+), but very well on T_1, T_2, T_3, T_4, T_5, and T_6 (++++) (Table 8.1).

8.7.2.4 *Trichophyton verrucosum* – This is cosmopolitan fungus that mainly affects cattle, although it can sporadically infect humans. The clinical diagnosis is particularly difficult, or even impossible, due to the absence of epidemiological data. It is clinically characterized as causing highly inflamed lesions of the scalp, glabrous skin, beard, and mustache. Microscopic examination of scalp samples reveals megaspore endothrix parasitism, without fluorescence under Wood's lamp. It grows very slowly, taking between 13 to 25 days to mature on Sabouraud agar.

Macroscopically, the colony has a texture that varies from glabrous to velvety, with a wrinkled or cerebriform surface. The pigmentation of the top varies from white to yellow ocher, which originated its old denomination *T. ochraceum*. The reverse is yellow, without pigment diffusion into the medium. Under optical microscopy, colonies grown on Sabouraud agar are frequently poorly developed, without macroconidia or microconidia. The most noteworthy feature of this species is the presence of long chains of large chlamydoconidia.

Physiologically, the species does not produce urease, nor is it able to perforate hair *in vitro* or produce pigment when grown on corneal agar. The majority of strains require thiamine to grow, and some need inositol. The growth intensity on *Trichophyton* culture media is as follows: no growth on T_1, T_5, T_6, and T_7; little (+/–) or no growth on T_2; slight growth on T_3(+); and excellent growth on T_4(++) (Table 8.1).

8.7.2.5 *Trichophyton violaceum* – It is an anthropophilic species involved in epidemics in schools and orphanages. It is mainly distributed in the Middle East, Europe, Africa, and Mexico. It causes scalp, skin, and nail lesions.

This species grows slowly on Sabouraud agar, forming colonies with violet color and glabrous texture with radial grooves. The reverse is also violet, without pigment diffusion. After several transfers, the colony loses its violet color and acquires a downy white mycelium.

It produces broad and tortuous hyphae, and generally presents no macroconidia nor microconidia. In colonies grown on thiamine-enriched media, pyriform macroconidia and microconidia can sometimes be observed.

8.8 Final considerations

In conclusion, considering the peculiarities of dermatophytes, polyphasic studies to obtain molecular, ecological, phenotypic, and life-cycle information are necessary to safely establish the species of these fungi, and, even then, their identification will be challenging.

References

Ahmadi, B., Mirhendi, H., Shidfar, M. R. et al. 2015. A comparative study on morphological versus molecular identification of dermatophyte isolates. *J Med Mycol* 25: 29–35.

Bouchara, J. P., Mignon, B., Chaturvedi, V. 2017. Dermatophytes and dermatophytoses: A thematic overview of state of the art, and the directions for future research and developments. *Mycopathologia* 182: 1–4.

Calderaro, A., Motta, F., Montecchini, S. et al. 2014. Identification of dermatophyte species after implementation of the in-house MALDI-TOF MS database. *Int J Mol Sci* 15: 16012–16024.

Coulibaly, O., L'Ollivier, C., Piarroux, R., Ranque, S. 2018. Epidemiology of human dermatophytoses in Africa. *Med Mycol* 56: 145–161.

de Hoog, G. S., Dukik, K., Monod, M. et al. 2017. Toward a novel multilocus phylogenetic taxonomy for the dermatophytes. *Mycopathologia* 182: 5–31.

de Hoog, G. S., Guarro, J., Gené, J., Figueras, M. J. 2000. *Atlas of Clinical Fungi*. 2nd ed, Centraalbureau voor Schimmelcultures/Universitat Rovira i Virgili.

Gnat, S., Nowakiwick, A., Ziólkowska, G., Troscianczyk, A., Majer-Dziedzic. B., Zieba, P. 2017. Evaluation of growth conditions and DNA extraction techniques used in the molecular analysis of dermatophytes. *J. Appl. Microbiol* 122: 1368–1379.

Gräser, Y., Monod, M., Bouchara, J. P. et al. 2018. New insights in dermatophyte research. *Med Mycol* 56(s1): 2–9.

Kupsch, C., Ohst, T., Pankewitz, F. et al. 2016. The agony of choice in dermatophyte diagnostics-performance of different molecular tests and culture in the detection of *Trichophyton rubrum* and *Trichophyton interdigitale*. *Clin Microbiol Infect* 22: 735.e11–7.

L'Ollivier, C., Ranque, S. 2017. MALDI-TOF based dermatophyte identification. *Mycopathologia* 182: 183 192.

Mochizuki, T., Takeda, K., Anzawa, K. 2015. Molecular markers useful for epidemiology of dermatophytoses. *J Dermatol* 42: 232–235.

Mochizuki, T., Takeda, K., Anzawa, K. 2017. Molecular markers useful for intraspecies subtyping and strain differentiation of dermatophytes. *Mycopathologia* 182: 57–65.

Nenoff, P., Kruger, C., Ginter-Hanselmayer, G., Tietz, H. J. 2014a. Mycology—An update. Part 1: Dermatomycoses: Causative agents, epidemiology and pathogenesis. *J Dtsch Dermatol Ges* 12: 188–209.

Nenoff, P., Kruger, C., Schaller, J., Ginter-Hanselmayer, G., Schulte-Beerbuhl, R., Tietz, H. J. 2014b. Mycology—An update. Part 2: Dermatomycoses: Clinical picture and diagnostics. *J Dtsch Dermatol Ges* 12:749–777.

Zhan, P., Liu, W. 2017. The changing face of the dermatophytic infections worldwide. *Mycopathologia* 182: 77–86.

9

Aspergillus spp.

Reginaldo Gonçalves de Lima-Neto,
Patrice Le Pape, and Rejane Pereira Neves

Contents

9.1 Introduction

The disease "asperqillosis" refers to allergy, respiratory tract invasion, cutaneous infection, or extrapulmonary dissemination caused by species of *Aspergillus*, mainly *A. fumigatus*, *A. flavus*, *A. niger*, and *A. terreus*. *Aspergillus* species are cosmopolitan and inhalation of spores is common. However, tissue invasion is not frequent, but may happen in the case of hosts which are immunosuppressed due to therapy for hematologic malignancies, receipt of hematopoietic cells, or solid organ transplantation (Shannon et al., 2010).

Aspergillus filamentous fungi are very common in all environments, so it is impossible to completely avoid breathing in some conidia. In people with healthy immune systems, this does not lead to infection. However, in people with debilitated immune systems, inhaling *Aspergillus* spores can cause lung or sinus infection, which can spread to other parts of the body. There are approximately 180 species of *Aspergillus*, but fewer than 40 of them are known to cause infections in humans (Barnes & Marr, 2006).

Indeed, *Aspergillus* mold is inescapable. Outdoors, it is found in decaying leaves and on plants, trees, and grain crops. Inside, the conidia thrive in air conditioners, heating ducts, and some foods. Everyday exposure to *Aspergillus* is rarely a problem for people with healthy immune systems because the defensive cells surround and destroy the spores. But a weakened immune system allows *Aspergillus* to take hold, invading the lungs, and, in the most serious cases, other parts of the body. Fortunately, aspergillosis is not contagious from person to person (Barnes & Marr, 2006; Shannon et al., 2010).

Aspergillus infection produces colonization, allergy, or invasive disease. Its manifestations are highly variable and should be considered based on the clinical history. Here, we review the diagnosis of invasive aspergillosis as well as diagnosis of other syndromes caused by *Aspergillus* clinical isolates, along with the risk factors, clinical manifestations, and current resistance mechanisms.

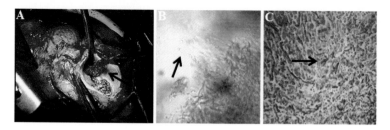

Figure 9.1   Invasive aspergillosis in a patient with aortic valve replacement. (A) *Aspergillus*-fungal vegetation in aortic artery (arrow). (B) Hyaline and septate hyphae with acute angle bifurcation (arrow). (C) Histological section showing septate hyphae with dichotomy by Grocott-Gomori methenamine silver staining (arrow).

9.2 Types and symptoms of aspergillosis

9.2.1 Invasive aspergillosis

This is a serious fungal infection that usually occurs in people who are already sick from other medical conditions, such as patients with weakened immune systems due to chemotherapy, organ transplant, or stem cell transplant. Invasive aspergillosis most commonly affects the lungs, but it can also spread to other parts of the body. The symptoms of invasive aspergillosis in the lungs can include fever and chills, chest and joint pain, cough, hemoptysis, shortness of breath, headaches or eye symptoms, nosebleed, facial swelling on one side, and skin lesions. The most serious complication occurs when the infection spreads rapidly from the lungs to the brain, heart (Figure 9.1), or kidneys. Untreated, this form of *Aspergillus* disease is usually fatal (Barnes & Marr, 2006).

9.2.2 Aspergilloma

A growth of tangled fungus fibers also called a "fungus ball" can develop in cavities, mainly in patients with pre-existing lung conditions, such as emphysema, tuberculosis, or advanced sarcoidosis. As the name suggests, it is a ball composed of *Aspergillus* hyphae, fibrin, mucus, and cellular debris found within a pulmonary cavity that grows in the lungs as part of chronic pulmonary aspergillosis or in the sinuses, but usually does not spread to other parts of the body. Indeed, aspergillomas generally develop in pre-existing pulmonary cavities (Denning et al., 2003; Lee et al., 2004).

Aspergilloma is usually a benign condition when it is present singly. It can remain stable over months, and may not produce symptoms or only produce a mild cough. But, over time, the underlying condition can worsen and possibly cause hemoptysis, wheezing, shortness of breath, and unintentional weight loss. In addition, aspergillomas can cause severe, and sometimes fatal, bleeding in the lungs (Barnes & Marr, 2006; Lee et al., 2004). However, aspergilloma does not produce many laboratory abnormalities that enable easy detection.

9.2.3 Chronic fibrosing pulmonary aspergillosis

This is a long-term (3 months or more) condition in which *Aspergillus* can cause cavities in the lungs and nodules. One or more fungal balls (aspergillomas) may also be present in the lungs. Subacute invasive pulmonary aspergillosis (formerly known as chronic necrotizing aspergillosis) is on the spectrum between chronic and acute forms of pulmonary aspergillosis. Disease duration longer than 3 months distinguishes chronic from acute and subacute pulmonary aspergillosis. Symptoms of chronic pulmonary aspergillosis include weight loss, cough, hemoptysis, shortness of breath, and fatigue (Denning et al., 2016; Denning et al., 2003).

9.2.4 Allergic aspergillosis

Allergic bronchopulmonary aspergillosis (ABPA) is characterized pathologically by mucoid impaction of the bronchi, eosinophilic pneumonia, and bronchocentric granulomatosis in addition to the histologic features of asthma. Areas of eosinophilic pneumonia are occasionally found, although this is not a major feature of the disease. Septate hyphae with acute dichotomous branching may be seen in the mucus-filled bronchial lumen, but fungi do not invade the mucosa. *Aspergillus* is cultured from the sputum in up to two-thirds of patients with ABPA, but hyphae may not be seen by direct microscopy. In these cases, *Aspergillus* causes inflammation in the lungs and allergy symptoms such as shortness of breath, coughing, and wheezing, but does not cause infection. Reported rates are higher in patients seen in asthma clinics and those admitted to the hospital with asthma exacerbation (Agarwal et al., 2013; Maturu & Agarwal, 2015).

Another allergic problem caused by *Aspergillus* is inflammation in the sinuses, causing symptoms of sinus infection such as stuffy nose, drainage that is possibly bloody, facial pain, headache, and reduced ability to smell. But it does not cause infection (Agarwal et al., 2013).

9.2.5 Cutaneous aspergillosis

Aspergillus enters the body through a break in the skin (e.g., after surgery or a burn wound) and causes infection, usually in people who have weakened immune systems. Cutaneous aspergillosis can also occur if invasive aspergillosis spreads to the skin from somewhere else in the body, such as the lungs (Barnes & Marr, 2006).

Onychomycosis due to *Aspergillus* species has become common in tropical regions, mainly among farmers and other field workers. Overall, the infection is chronic, with advanced disease involving several nails, causing distorted architecture of the nails of the hands or feet (Barnes & Marr, 2006).

9.3 Aspergillosis risk factors and prevention

The risk to develop aspergillosis depends on overall health and the level of exposure to conidia. In general, the factors that most predispose patients to infection are weakened immune system, low white blood cell level, lung cavities, asthma or cystic fibrosis, and long-term corticosteroid therapy (Agarwal et al., 2013).

Invasive aspergillosis affects patients taking immune-suppressing drugs after undergoing transplant surgery, especially bone marrow or stem cell transplants, or patients who have leukemia and other malignant hematological diseases. Patients in the later stages of AIDS also may be at increased risk (Barnes & Marr, 2006).

Aspergillomas usually affect patients who have healed air spaces (cavities) in their lungs and are at higher risk of developing a mass of tangled fungus fibers. Inflammatory illnesses like tuberculosis or sarcoidosis, as well as radiation, may damage the lungs, producing these cavities (Barnes & Marr, 2006; Lee et al., 2004).

Chronic pulmonary aspergillosis typically occurs in people who have other lung diseases, including chronic obstructive pulmonary disease, besides non-cancerous and inflammatory illnesses like tuberculosis or sarcoidosis (Schweer et al., 2014).

Allergic response to *Aspergillus* is most common in people with asthma and cystic fibrosis, especially those whose lung problems are longstanding or hard to control (Agarwal, 2014; Maturu & Agarwal, 2015).

It is almost impossible to control exposure to *Aspergillus* spores, but transplant patients and those undergoing chemotherapy should to stay away from the obvious sources of mold, such as construction sites, compost piles, and stored grain. In addition, wearing a face mask to decrease exposure to airborne infectious agents is recommended (Barnes & Marr, 2006).

9.4 Approach to diagnosis

Definitive diagnosis of invasive aspergillosis or chronic necrotizing *Aspergillus* pneumonia depends on identification of the organism in tissue samples. *Aspergillus* spp. culture in association with the presence detected by direct microscopy from bronchoalveolar lavage (BAL) or sputum

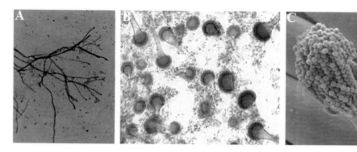

Figure 9.2   Laboratory findings for Aspergillosis by direct microscopic examination and culture. (A) Grocott silver stain coloration revealed septate branching hyphae of *A. fumigatus*. (B) Heads of *Aspergillus fumigatus* revealed by lactophenol cotton blue staining. (C) Head of *Aspergillus fumigatus* observed by scanning electron microscopy.

of septate and dichotomous hyphae provide definitive evidence of this mycosis (Figure 9.2). It should be noted that only the visualization of septate and dichotomous hyphae in direct microscopy without positive culture for *Aspergillus*, can be correspond to an hyphomycosis only. Histopathological examination showing tissue damage by hyphae using Gomori methenamine silver stain or calcofluor are valuable for diagnosis also, but biopsy is frequently not feasible due to the risks of complications (Barnes & Marr, 2006).

Caution is necessary with patients who have clinical and radiological findings that are suggestive of invasive aspergillosis. However, when both the serum galactomannan assay and fungal stain and culture of the sputum are negative, bronchoscopy with BAL should be performed. In addition, lung fragments should be obtained if feasible. Indeed, a positive fungal stain from BAL, sputum or lung biopsy, and culture should prompt therapy of hosts who are neutropenic or are being treated for cancer. This is especially important after bone marrow transplantation in case of positive *Aspergillus* culture from sputum, since this has 95% positive predictive value for invasive aspergillosis. Nevertheless, a negative culture from BAL or sputum does not rule out pulmonary aspergillosis. Some 45%–60% of patients present *Aspergillus* cultured from BAL eventually found by biopsy or autopsy, and only 8%–30% of patients present *Aspergillus* cultured from sputum (Sampsonas et al., 2011; Shannon et al., 2010).

The use of unspecific serum biomarkers like beta-D-glucan, a pan-fungal test, or *Aspergillus*-specific biomarkers as galactomannan are other rational steps to establish the diagnosis of invasive aspergillosis involving the choice of noninvasive specimens. When BAL is obtained, a sample should be sent to detect galactomannan, a major component of the *Aspergillus* cell wall (Luong et al., 2011). Patients who are at high risk should be also monitored closely for invasive *Aspergillus* infection by evaluating serum galactomannan levels weekly. Furthermore, the evaluation of serum galactomannan found that charting early trends during the first 15 days of antifungal treatment may be helpful to anticipate good clinical response. A reduction in galactomannan levels between baseline and the first 7 days predicted clinical outcomes. A meta-analysis and systematic review determined that the measurement of BAL-galactomannan levels may help in diagnosing invasive aspergillosis early, since BAL fluid testing is carried out in the suspicious area (Sampsonas et al., 2011; Shannon et al., 2010).

Another assay useful for diagnosis of invasive aspergillosis in high-risk patients is real-time polymerase chain reaction (PCR). Luong et al. (2011) concluded that pan-*Aspergillus* PCR associated with BAL galactomannan samples from lung transplant recipients was 97% specific and 93% sensitive for invasive pulmonary aspergillosis. Moreover, species-specific real-time PCR testing for *A. fumigatus* and *A. terreus* can be used to diagnose the common *A. fumigatus* and the amphotericin B-resistant *A. terreus*.

Morio et al. (2018) showed that a PCR-based strategy is of major interest, offering increased sensitivity compared with mycological cultures, identifying 78.9% of *Aspergillus fumigattus* from clinical samples compared with only 32% when using plate cultures as gold standard in a cohort of 137 French patients with fungal rhinosinusitis. In this study, each sample was

Table 9.1　Target Regions and Primers Used for Amplification and Sequencing of *Aspergillus* Species according to Samson et al. (2014)

Target Region	Primers
Beta-tubulin	Bt2a (5′-GGT AAC CAA ATC GGT GCT GCT TTC-3′) Bt2b (5′-ACC CTC AGT GTA GTG ACC CTT GGC-3′)
Calmodulin	Cmd5 (5′-CCG AGT ACA AGG ARG CCT TC-3′) Cmd6 (5′-CCG ATR GAG GTC ATR ACG TGG-3′)

subjected to external lysis with 300 µL of tissue lysis buffer for 15 min at room temperature, and then automated DNA extraction was performed with 10 µL of the commercial four-plex PCR assay from Qiagen. Table 9.1 shows the common target regions and primers that are used for amplification and sequencing of *Aspergillus* strains.

Rapid laboratory diagnosis of the main *Aspergillus* clinical species can be achieved by using proteomic techniques by matrix-assisted laser desorption and ionization time-of-flight mass spectrometry (MALDI-TOF-MS). There are 19 main spectra projections present in the Biotyper library (MALDI Biotyper 3.1, Bruker Daltonics, Bremen, Germany), one being *Aspergillus brasiliensis*, 2 *A. flavus*, 6 *A. fumigatus*, 6 *A. niger*, 2 *A. terreus*, 1 *A. versicolor*, and 1 *A. thermomutatus* [anamorph] (*Neosartorya pseudofischeri* [teleomorph]) (Bruker Daltonics, 2017). The cultivation procedure for filamentous fungi, such as *Aspergillus*, and the formic acid extraction protocol are described below:

1. Add a small amount of biological material to tubes containing at least 5 mL of Sabouraud liquid broth and close the tube.
2. Rotate the rotator to shake overhead until enough biological material is observed (usually 24 h for *Aspergillus* species).
3. Remove the cultivation tubes from the rotator, place them standing up and wait for few minutes (allowing the filamentous fungi to settle to the bottom of the tubes).
4. Harvest up to 1.5 mL from the sediment and transfer it to a microcentrifuge tube.
5. Centrifuge at full speed (13,000 rpm) for 2 min.
6. Carefully remove the supernatant.
7. Add 1 mL of water to the pellet and vortex for 1 min.
8. Centrifuge at full speed (13,000 upm) for 2 min.
9. Carefully remove the supernatant again and repeat washing and vortexing once.
10. Suspend the pellet in 300 µL of water, add 900 µL of ethanol, and vortex it.
11. Centrifuge at full speed (13,000 upm) for 2 min.
12. Remove the supernatant carefully by pipetting (avoid decanting), centrifuge briefly, and remove the residual ethanol completely.
13. Dry the pellet completely using a Speedvac or drying at 37°C.
14. Resuspend pellet in a certain amount of 70% formic acid (a very small pellet will require 10–20 µL and a big pellet could require up to 100 µL of formic acid).
15. Add the same volume of acetonitryl to the tube and mix it carefully.
16. Centrifuge at full speed (13,000 upm) for 2 min.
17. Add 1 µL of supernatant on the MALDI target plate and allow to air dry.
18. Overlay with 1 µL of matrix solution (&-cyano-4-hydroxycinnamic acid) to cover each sample spot; air dry (ensure target is completely dry before it is inserted into the MALDI device).

9.5　Resistance in *Aspergillus* spp.

The emergence of triazole resistance is increasingly reported around the world in the past decade, becoming a worrying public health problem. In a patient, this resistance is acquired through two distinct epidemiological routes. The first resistance cases were described following long-term treatment with medical azoles (Denning et al., 1997; Chryssanthou, 1997; Verweij et al., 2016). More recently, contamination of azole-naive patients by already resistant isolates from the environment has been reported, linked to the use of azole fungicides in agriculture, horticulture, and wood preservation (Alvarez-Moreno et al., 2017; Snelders et al., 2008; Verweij et al., 2016). This second route of acquisition of resistance was initially proposed in the Netherlands and then

10
Mucorales

Rejane Pereira Neves, André Luiz Cabral Monteiro de
Azevedo Santiago, and Reginaldo Gonçalves de Lima-Neto

Contents

10.1 Introduction (epidemiology and common mucormycosis agents)

The order Mucorales Dumort belongs to the family Mucoraceae Dumort., subphylum Mucoromycotina Benny and phylum Mucoromycota Doweld, comprising the largest number of species (more than 205 spp.) in the phylum to which it belongs (Kirk et al., 2008; Spatafora et al., 2016). Species of this order are commonly found in soil as well as in excrement of herbivores and rodents, although some taxa are commonly isolated from decaying fruits and stored food (Hoffmann et al., 2013). In general, mucoralean fungi differ from other fungi by producing zygospores (a sexual spore type), as well as asexual structures and spores, such as sporangiophores, sporangiospores, merosporangiophores, merosporangiospores, sporangioles, chlamydospores, and, rarely, artrospores (Alexopoulos et al., 1996). Mucorales species exhibit cenocytic hyphae, although septa delimiting reproductive structures can be visualized in several taxa, as well as irregularly spaced septa, which can be observed in aged colonies (de Souza et al., 2017). In general, these species produce a fast-growing mycelium with intensely branched hyphae and a cottony appearance. The coloration of colonies varies according to the species, environmental temperature, and colony age (Trufem et al., 2006).

The ability of mucoralean fungi to adapt to adverse environmental conditions as a result of physiological, biochemical, and genetic mechanisms is reflected in specific strategies for development, maturation, differentiation and survival, allowing the pioneer colonization of substrates, since these fungi prefer to feed on less complex sugars (Richardson, 2009), although some species are capable of degrading complex sugars, such as pectin and hemicelluloses (Domsch et al., 2007). Being dependent on soluble carbohydrates, these fungi need to utilize these sugars before they are used by other microorganisms. Therefore, rapid sporulation and accelerated mycelial growth, even in nutrient-poor media, are hallmarks of most taxa of this order. It is common for a Petri dish with 9 cm diameter, inoculated with Mucorales spores and incubated at 25°C to 28°C, to be completely filled by the fungus in less than 48 hours.

Several species of Mucorales are known for their ability to produce enzymes, such as amylases and inulinases, as well as lactic and fumaric acids (Hoffmann et al., 2013), with high industrial importance due to use in the production of fermented foods, mainly in Asian countries. In agriculture, some species, such as *Choanephora cucurbitarum* (Berk & Ravenel) Thaxt. and *Rhizopus stolonifer* (Ehrenb.) Vuill. play an important role as plant pathogens (Benny et al., 2005; Bassey et al., 2018), whereas other species of Mucorales are able to cause infections in human patients (Bouza et al., 2006; Sipsas et al., 2018).

Mucormycosis is a rare infection caused by fungi of the Mucorales order, mainly affecting immunocompromised patients, with scarce reports in healthy individuals (Schwartze et al., 2014; Vaezi et al., 2016). The most common agents of mucormycosis are those of the genera *Rhizopus*, *Lichtheimia*, and *Mucor*, with *Rhizopus* species being the most prevalent opportunistic pathogens (Vaezi et al., 2016), whereas species of *Rhizomucor*, *Saksenaea*, *Cunninghamella*, *Apophysomyces*, and *Syncephalastrum* are less common agents (Bouza et al., 2006; Petrikkos et al., 2012). The incidence of mucormycosis is approximately 1.7 cases per 1,000,000 inhabitants per year, and the major risk factors are uncontrolled diabetes mellitus, hematological malignancy, use of corticosteroids, burns or other trauma, and organ transplant (Cunha et al., 2011; Ibrahim et al., 2003; Song et al., 2017; Spellberg et al., 2005; Sugar, 2005).

Mucormycosis infections are characterized by angioinvasion that causes vessel thrombosis and tissue necrosis, and the disease may have rhinocerebral, pulmonary, cutaneous, gastrointestinal, or disseminated forms. Each of these categories can be related to specific defects in the host's defense (Spellberg et al., 2005).

This chapter summarizes the knowledge about predisposing factors, pathogenesis, clinical presentation, diagnosis, and treatment of mucormycosis.

10.2 Predisposing factors and pathogenesis

The major predisposing factors are intravenous drug use; uncontrolled diabetes mellitus, especially with ketoacidosis; hematological disorders like leukemia and lymphoma; exposure to high iron content; infection by human immunodeficiency virus; immunosuppression with severe neutropenia; solid organ transplantation and bone marrow transplantation; treatment with deferoxamine; extensive burns or major trauma; and long-term antifungal prophylaxis with voriconazole as well as caspofungin. The most frequent predisposing factor varies among the published articles. However, diabetes, hematological disorders, and immunosuppression have been shown to be more prevalent (Petrikkos et al., 2012; Roden et al., 2005; Vaezi et al., 2016). In addition, there are several reports of mucormycosis after traumatic inoculation of contaminated soil (Adam et al., 1994).

Mucormycosis presents a higher incidence in patients with uncontrolled diabetes (mostly in patients with ketoacidosis), since acid pH promotes the dissociation of Fe^+ from transferritin, increasing the free fraction to be incorporated in the fungus. It also occurs in patients treated with iron chelators, since these are necessary for the development and virulence of Mucorales fungi. Therefore, the plasma level of $Fe+$ ions is usually higher in mucormycosis (Ibrahim et al., 2008, 2012).

According to the initial involvement by the fungus, the clinical form of mucormycosis develops. Thus, if the primary lesion occurs on the face, the fungus typically spreads to the orbit and brain, characterizing the clinical rhinocerebral form (Singh et al., 2013). In addition, these fungi have a remarkable ability to invade vessel walls, culminating in thrombosis, causing localized necrosis. This facilitates the penetration into the fine bones that delimit the nasal sinuses, orbit, and base of the skull, such as the cryptic laminae of the ethmoid. The results are abscesses and meningoencephalitis, often accompanied by cerebral infarctions secondary to thrombosis (Gamaletsou et al., 2012).

Mucoralean fungi have a strong tropism for arteries and penetrate the internal elastic lamina, promoting endothelial damage, thrombosis, and infarction. Indeed, it is believed that acidosis facilitates the invasion of blood vessel walls by these fungi due these infectious agents' presence in the keto-reductase pathway, which is activated in acidic pH caused by uncontrolled diabetes (Abdollahi et al., 2016).

10.3 Clinical features

Mucormycosis can occur in many parts of the human body, such as the head and neck, lung, skin, gastrointestinal tract, central nervous system, and primarily in the nasal cavity (Petrikkos et al., 2012). Inhalation of sporangiospores—the Mucorales fungal spores—which are present in the environment, is the principle mode of acquisition and colonization of the spores in the nasal/sinus mucosa. Following inhalation, fungal elements are deposited in neighboring areas including orbit, brain, cavernous sinus, and pulmonary alveoli, where they can precipitate allergic sinusitis and interstitial pneumonitis in immunocompetent hosts. Invasive sinus infection and pneumonia occur in immunocompromised patients (Gamaletsou et al., 2012). Sinus disease by Mucorales is the most commonly reported presentation and may be localized or extended to the orbit and/or brain. Necrotic ulcers on nasal mucosa or turbinates are revealed by nasal endoscopy. Medical and surgical emergencies occur when infection does not remain contained within paranasal sinuses and progresses to sino-orbital and rhinocerebral manifestation. This progression has a constellation of clinical features (Abdollahi et al., 2016).

Usually, sinus mucormycosis causes nasal congestion, dark blood-tinged rhinorrhea or epistaxis, sinus tenderness, retro-orbital headache, fever, and malaise. More advanced sinus infection may cause facial or periorbital swelling and numbness, blurred vision, lacrimation, chemosis, diplopia, proptosis, and loss of vision in the affected eye (Pelton et al., 2001). Infection also can extend to adjacent bone and ultimately to the skull base. Once penetrated within the orbital compartment, the fungal elements can invade other muscles, besides extending posteriorly to the optic foramen, where the ophthalmic artery, ophthalmic nerve, and optic nerve are threatened by invasion, edema, inflammation, and necrosis (Deshazo, 2009).

Progression to the central nervous system occurs via the optic nerve or from the ethmoid sinuses by way of the cavernous sinus. Abnormal mentation often signifies cerebral involvement (Abdollahi et al., 2016).

Cutaneous mucormycosis usually occurs in damaged skin. Penetrating trauma, dressings, and burns are the most common reasons. The most common manifestations are erythema or induration of the skin and necrotic eschar.

Muscle, tendon, or bone may be affected due invasive extension from cutaneous tissues (Cheng et al., 2017).

Gastrointestinal mucormycosis accounts for about 7% of all cases, but reaches a mortality rate of up to 85% (Choi et al., 2012) and is most commonly observed in the stomach, followed by the small intestine and colon (Spellberg, 2012). The clinical manifestation may be evidenced as ischemic bowel disease or gastrointestinal bleeding (Sun et al., 2017). In addition, the pediatric literature has reported a number of cases of necrotizing enterocolitis caused by Mucorales (Vallabhaneni et al., 2015).

Although gastrointestinal mucormycosis is a rare disease in immunocompetent hosts, 10 cases have been reported in the literature since 2000, of which 4 cases occurred in the stomach, 1 in the jejunum, 5 in the colon (including one case occurring in both the stomach and colon), and 1 in the abdominal cavity after a trauma (Sun et al., 2017).

10.4 Diagnosis

Laboratory diagnosis is indispensable to prescribe adequate treatment for a favorable clinical outcome, since mucormycosis is often fatal when undiagnosed and/or not treated early (Brettholz and Mccauley, 2018).

A combination of host factors, clinical manifestations, and radiological findings may help with early diagnosis of mucormycosis, but definitive diagnosis relies on: (a) direct microscopy, (b) culture of tissue and aspirated samples, (c) histopathology, and (d) molecular diagnosis in tissue samples. Laboratorial diagnosis of mucormycosis remains a challenge due to the delicacy, fragility, and low viability of fungal elements, which often causes false negative culture results (Dannaoui, 2009). Advanced molecular and other non–culture based approaches are promising and may be used to complement the conventional methods.

Deep-tissue biopsy of the involved site is usually cumbersome because the patients are often too unstable to undergo invasive procedures. This biological sample is usually taken by functional endoscopic sinus surgery or needle aspiration (De Pauw et al., 2018).

The direct examination by microscopy of tissue biopsy or aspirated material using 10%–20% potassium hydroxide (KOH), with observation of poorly septated hyphae and ribbon-like hyphae

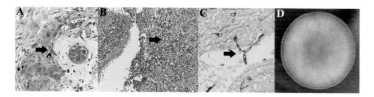

Figure 10.1   Microscopic morphology of Mucorales in tissue section stained with PAS (A), HE (B), and Grocott-Gomori (C). Mucorales produce wide poorly septate hyphae with wide angle (>90°C) in tissues (black arrows; A, C). A bubbly tissue appearance may be seen in areas where the hyphae were cross-sectioned (black arrow; B). (D) *Mucor irregularies* on Sabouraud dextrose agar. Colony with cottony aspect expanded to the lid. Figures (B) and (C) were kindly provided by Dr. Sami Gadelha and Dr. José Telmo Valença Junior (Federal University of Ceará, Brazil).

Figure 10.2   Mucormycosis agents. (A) *Rhizopus arrhizus* var. *arrhizus*: sporangiophores long, in groups, frequently reaching 1.5 µm in length, with columellae and rhizoids (arrow); (B) *Cunninghamella echinulata*: branched sporophore with sporangiola, long and short branches in the same sporophore; (C, D) *Mucor racemosus* f. *racemosus*: sporangiophore with a non-apophysate sporangia (C) and columellae (D), rhizoids absent; (E) *Rhizopus microsporus*: sporangiophores short, in pairs, rarely reaching 1 µm in length, with sporangia, columellae, and rhizoids (arrow); (F, G) *Lichtheimia ramosa*: branched sporangiophore with sporangia (F), columellae oval, globose, elongate, and spatulate (G); (H) *Syncephalastrum racemosum*: meroraporangiophore with merosporangia and merospores.

branching at wide angles associated with tissue damage, is strong evidence of mucormycosis, so a biological sample should be cultured. In the case of patients with histopathological evidence of mucormycosis in tissue biopsy presenting the same structures observed in direct examination (Figure 10.1A–C), clinical samples should also be cultured. However, several researchers argue that the gold standard for diagnosis remains pathologic findings of a tissue biopsy (Abdollahi et al., 2016; Cheng et al., 2017; Dannaoui, 2009; Sun et al., 2017). Furthermore, the rapid growth of grey fluffy colonies resembling cotton candy on Sabouraud dextrose agar (SDA) plus 2 g/L chloramphenicol (Figure 10.1D) incubated in duplicate at 30°C and 37°C, showing typical sporophores (Figure 10.2), corroborates the mucormycosis diagnosis. Some researchers consider that the growth of pure and similar colonies of Mucorales on more than one culture media is significant (Abdollahi et al., 2016; Haghani et al., 2015).

Imaging is unspecific, but is very useful and recommended to evaluate the extent of the disease. Computed tomography supports evidence of the thickening of the mucosa and turbidity in the sinuses, besides fluid levels, bone destruction, and osteomyelitis. Moreover, magnetic resonance imaging may be used to assess the spread of infection to the orbit and cranial area. Radiography can reveal opacification of the paranasal sinuses during mucormycosis (Abdollahi et al., 2016).

In 2009, an important review summarized the molecular approaches available for identification of Mucorales and diagnosis of mucormycosis (Table 10.1). This study was conducted according to the standards of the International Society of Human and Animal Mycology (ISHAM) Working Group on Fungal Molecular Identification and proposed the use of ITS sequencing as a first-line strategy for Mucorales identification in tissues or in culture specimens (Dannaoui, 2009).

Finally, matrix-assisted laser desorption and ionization time-of-flight mass spectrometry (MALDI-TOF-MS), which has been utilized for identification of various microorganisms and reported to

Table 10.1 Molecular Methods and Targets Available for Species Identification in Mucorales Culture, Fresh or Frozen Tissues

Sample	Molecular Method	Target Region	Species
Culture	PCR + sequencing	28S	Several species
	PCR + RFLP	18S	Several species
	MicroSeq	28S	Several species
	PCR + sequencing	ITS	Several species
	Real-time PCR	cyt b	Several species
	PCR + sequencing	ITS	Several species
	Multiplex PCR	ITS	*Rhizopus* species
	PCR + sequencing	ITS	*Rhizopus oryzae*
	PCR + sequencing	ftr1	*Rhizopus* species
	PCR + RFLP	ITS	*Apophysomyces elegans*
Fresh/frozen host tissues	PCR + sequencing	ITS	*Rhizopus oryzae, Apophysomyces elegans*
	PCR + sequencing	28S	*Cunninghamella bertholletiae*
	PCR + RFLP	18S	*Rhizopus microsporus*
	PCR + RFLP	18S	*Mycocladus corymbifer, R. microsporus*
	PCR + sequencing	ITS	*Rhizomucor pusillus*
	PCR + sequencing	ITS	*Saksenaea vasiformis*
	Real-time PCR	cyt b	*Apophysomyces elegans*

Abbreviations: 28S, large subunit ribosomal DNA; 18S, small subunit ribosomal DNA; ITS, ribosomal DNA internal transcribed spacer; cyt b, cytochrome b gene; ftr1, high-affinity iron permease 1 gene; PCR, polymerase chain reaction; RFLP, restricted fragment length polymorphism.

be a rapid and reliable diagnostic tool to identify environmental and clinically important yeasts, has not been effective regarding Mucorales species. There are only two Main Spectra Projections present in the Biotyper library (MALDI Biotyper 3.1, Bruker Daltonics, Bremen, Germany), one of *Mucor circinelloides* and the other of *Rhizopus microsporus*.

10.5 Therapy

The therapeutic strategies against mucormycosis include antifungal therapy, surgery, and reduction of immunosuppression. Patients often receive both antifungals and surgery therapy. Aggressive surgical debridement associated with anti-mucormycosis drugs like amphotericin B/posaconazole is a better choice than antifungal therapy alone. Treatment with liposomal amphotericin B or deoxycholate amphotericin B is also common (Abdollahi et al., 2016; Cheng et al., 2017; Sun et al., 2017).

References

Abdollahi, A., Shokohi, T., Amirrajab, N. et al. 2016. Clinical features, diagnosis, and outcomes of rhino-orbito-cerebral mucormycosis—A retrospective analysis. *Curr Med Mycol* 2: 15–23.

Adam, R. D., Hunter, G., DiTomasso, J., Comerci Jr., G. 1994. Mucormycosis: Emerging prominence of cutaneous infections. *Clin Infect Dis* 19: 67–76.

Alexopoulos, C. J., Mims, C. W., Blackwell, M. 1996. *Introductory Mycology.* 4th ed. New York: John Wiley & Sons.

Bassey, A. R., Ubi, G. M., Patrick, A. 2018. Effect of *Choanephora Cucurbitarum* on the morphology of some plants in the Malvaceae family in Calabar, Cross River State, Nigeria. *IOSR J Pharm Biol Sci* 13(1): 53–59.

Benny, G. L. 2005. Zygomycetes. Pubished on the internet at http://www.zygomycetes.org.

Bouza, E., Muñoz, P., Guinea, J. 2006. Mucormycosis: An emerging disease? *Clin Microbiol Infect* 12: 7–23.

Brettholz, A. M., Mccauley, S. O. 2018. Mucormycosis: Early identification of a deadly fungus. *J Pediatr Oncol Nurs* 35: 257–266.

Cheng, W., Wang, G., Yang, M. et al. 2017. Cutaneous mucormycosis in a patient with lupus nephritis. A case report and review of literature. *Medicine* 96: 42–45.

Choi, H. L., Shin, Y. M., Lee, K. M. et al. 2012. Bowel infarction due to intestinal mucormycosis in an immunocompetent patient. *J Korean Surg Soc* 83: 325–329.

Cunha, M., Nery, A., Lima, F., Junior, J. D., Neto, J. M., Calado, N. 2011. Rhinocerebral zygomycosis in a diabetic patient. *Ver Soc Bras Med Trop* 44: 257–259.

Dannaoui, E. 2009. Molecular tools for identification of Zygomycetes and the diagnosis of zygomycosis. *Clin Microbiol Infect* 15: 66–70.

De Pauw, B., Walsh, T. J., Donnelly, J. P. et al. 2008. Revised definitions of invasive fungal disease from the European Organization for Research and Treatment of Cancer/ Invasive Fungal Infections Cooperative Group and the National Institute of Allergy and Infectious Diseases Mycoses Study Group (EORTC/MSG) Consensus Group. *Clin Infect Dis* 46: 1813–1821.

De Souza, C. A. F., Lima, D. X., Gurgel, L. M. S., Santiago, A. L. C. M. A. 2017. *Coprophilous* Mucorales (ex Zygomycota) from three areas in the semi-arid of Pernambuco, Brazil. *Braz J Microbiol* 4: 79–86.

Deshazo, R. D. 2009. Syndromes of invasive fungal sinusitis. *Med Mycol* 47: S309–14.

Domsch, K. H., Gams, W., Anderson, T. H. 2007. *Compendium of Soil Fungi.* San Francisco: IHW-Verlag.

Freire, K. T. L. S., Araújo, T. R., Bezerra, J. D. P. et al. 2015. Fungos Endofíticos de *Opuntia ficus-indica* (l.) Mill. (Cactaceae) sadia e infestada por *Dactylopius opuntiae* (Cockerell, 1896) (Hemiptera: Dactylopiidae). *Gaia Scientia* 9: 104–110.

Gamaletsou, M. N., Sipsas, N. V., Roilides, E., Walsh, T. J. 2012. Rhino-orbital-cerebral mucormycosis. *Curr Infect Dis Rep* 14: 423–434.

Haghani, I., Amirinia, F., Nowroozpoor-Dailami, K., Shokohi, T. 2015. Detection of fungi by conventional methods and semi-nested PCR in patients with presumed fungal keratitis. *Curr Med Mycol* 1: 31–38.

Hoffmann, K., Pawłowska, J., Walther, G. et al. 2013. The family structure of the Mucorales: A synoptic revision based on comprehensive multigene-genealogies. *Persoonia* 30: 57–76.

Ibrahim, A. S., Edwards, J. E., Filler, S. G. 2003. Zygomycosis. In: Dismukes W.E., Pappas P.G., Sobel J.D. (eds.) *Clinical Mycology*. New York, NY: Oxford University Press.

Ibrahim, A. S., Spellberg, B., Edwards, J. 2008. Iron acquisition: A novel perspective on mucormycosis pathogenesis and treatment. *Curr Opin Infect Dis* 21: 620–625.

Ibrahim, A. S., Spellberg, B., Walsh, T. J., Kontoyiannis, D. P. 2012. Pathogenesis of mucormycosis. *Clin Infect Dis* 54: 16–22.

Kirk, P. M., Cannon, P. F., Minter, D. W., Stalpers, J. A. 2008. *Dictionary of the Fungi*. 10th ed. Wallingford, UK: CAB International.

Pelton, R. W., Peterson, E. A., Patel, B. C., Davis, K. 2001. Successful treatment of rhino-orbital mucormycosis without exentration: The use of multiple treatment modalities. *Ophthal Plast Rconstr Surg* 17: 62–66.

Petrikkos, G., Skiada, A., Lortholary, O., Roilides, E., Walsh, T. J., Kontoyiannia, D. P. 2012. Epidemiology and clinical manifestations of mucormycosis. *Clin Infect Dis* 54: 23–34.

Richardson, M. 2009. The ecology of the Zygomycetes and its impact on environmental exposure. *Clinl Microbiol Infec* 15: 2–9.

Roden, M. M., Zaoutis, T. E., Buchanan, W. L. et al. 2005. Epidemiology and outcome of zygomycosis: A review of 929 reported cases. *Clin Infect Dis* 41: 634–653.

Schwartze, V. U., Santiago, A. L. C. M. A., Jacobsen, I. D., Voigt, K. 2014. The pathogenic potential of the *Lichtheimia* genus revisited: *Lichtheimia brasiliensis* is a novel, non-pathogenic species. *Mycoses* 57: 128–131.

Singh, V., Singh, M., Joshi, C., Sangwan, J. 2013. Rhinocerebral mucormycosis in a patient with type 1 diabetes presenting as toothache: A case report from Himalayan region of India. *BMJ Case Rep* pii: bcr2013200811. Doi: 10.1136/bcr-2013-200811

Sipsas, N. V., Gamaletsou, M. N., Anastasopoulou, A., Kotoyiannis, D. P. 2018. Therapy of mucormycosis. *J Fungi* 4(3): 90.

Song, Y., Qiao, J., Giovanni, G. et al. 2017. Mucormycosis in renal transplant recipients: Review of 174 reported cases. *BMC Infect Dis* 17: 283–288.

Spatafora, J. W., Chang, Y., Benny, G. L. et al. 2016. A phylum-level phylogenetic classification of zygomycete fungi based on genome-scale data. *Mycologia* 108: 1028–1046.

Spellberg, B. 2012. Gastrointestinal mucormycosis: An evolving disease. *Gastroenterol Hepatol* 8: 140–142.

Spellberg, B., Edwards, J., Jr., Ibrahim, A. 2005. Novel perspectives on mucormycosis: Pathophysiology, presentation, and management. *Clin Microbiol Rev* 8: 556–569.

Sugar, A. M. 2005. Agents of mucormycosis and related species. In: Mandell G.L., Bennett J.E., Dolin R. (eds.) *Principles and Practice of Infectious Diseases*. 6th ed. Philadelphia: Elsevier.

Sun, M., Hou, X., Wang, X., Chen, G., Zhao, Y. 2017. Gastrointestinal mucormycosis of the jejunum in an immunocompetent patient. A case report. *Medicine* 96: e6360.

Trufem, S. F. B., Maia, L. C., Souza-Motta, C. M., Santiago, A. L. C. M. A. 2006. Filo Zygomycota. In: Giulietti A.M. (ed.) *Diversidade e caracterização dos fungos do semi-árido brasileiro*. Recife, Brazil: Instituto do Milênio do Semi-árido, pp. 97–107.

Vaezi, A., Moazeni, M., Rahimi, M. T., de Hoog, S., Badali, H. 2016. Mucormycosis in Iran: A systematic review. *Mycoses* 59: 402–415.

Vallabhaneni, S., Walker, T. A., Lockhart. S. R. et al. 2015. Centers for Disease Control and Prevention. Notes from the field: Fatal gastrointestinal mucormycosis in a premature infant associated with a contaminated dietary supplement: Connecticut, 2014. *MMWR Morb Mortal Wkly Rep* 64: 155–156.

11
Sporothrix spp.

Anderson Messias Rodrigues, Rosane Orofino-Costa,
and Zoilo Pires de Camargo

Contents

The emergence of newly identified fungal pathogens such as *S. brasiliensis* and *S. globosa* and the reemergence of previously uncommon animal sporotrichosis have raised several concerns about laboratory diagnosis. Methods to recognize and diagnose classical and new agents are described in depth here, covering the latest taxonomic changes among clinically relevant *Sporothrix* species.

11.1 Introduction

Sporotrichosis is a subacute or chronic infection caused by the classical agent *Sporothrix schenckii* and the newly described species *S. brasiliensis*, *S. globosa*, and *S. luriei* (Marimon et al., 2007). Less frequent agents are classified in an environmental clade and include *S. chilensis* (Rodrigues et al., 2016b), *S. pallida* (Morrison et al., 2013), *S. mexicana* (Rodrigues et al., 2013b), and *S. stenoceras* (Dixon et al., 1991; Mayorga et al., 1978) (Figure 11.1).

Sporothrix spp. exists in nature, where it has been isolated from soil, decaying wood, *Sphagnum* moss, reed leaves, corn stalks, hay, and plant debris (Rodrigues et al., 2014a). Medically relevant *Sporothrix* spp. occurs in ecological niches of soil with wide ranges of temperature (6°C–28°C) and moisture (37.5%–99%) (Ramírez-Soto et al., 2018). Nevertheless, the ecological features underpinning population fluctuation are still enigmatic (Zhang et al., 2015).

After traumatic inoculation of *Sporothrix* propagules into a warm-blooded host, sporotrichosis may develop as a subacute or chronic infection, mainly affecting the skin and subcutaneous tissues, usually causing nodular lesions, which soften and break down to form indolent ulcers, with regional lymph nodes serving for local lymphatic drainage. Systemic involvement may occur (Silva-Vergara et al., 2012). Depending on factors such as inoculum size and host immunity, the incubation period is highly variable, with a mean of approximately 3 weeks (Barros et al., 2011).

Infections have been reported in people ranging from young children to adults and the elderly, with no differences related to sex, and all races seem to be equally susceptible to *Sporothrix* spp. People working with soil and plants compose a risk group for acquisition of the disease. As an

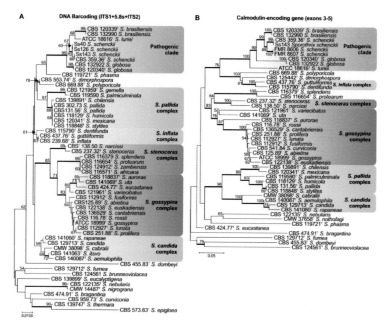

Figure 11.1   Phylogenetic tree generated by using partial nucleotide sequences of the (A) rDNA operon (ITS1+5.8s+ITS2) or (B) the calmodulin encoding gene. Neighbor-joining (NJ) and maximum likelihood (ML) analyses were performed using a Kimura 2-parameter model. The percentage of replicate trees in which the associated taxa clustered together in the bootstrap test (1000 replicates) is shown next to the branches (NJ/ML). Bar = total nucleotide differences between taxa. Reference sequences were retrieved from Genbank and published previously (de Beer, Z. W. et al. 2016. *Stud Mycol* 83: 165–191; Zhou, X., et al. 2014. *Fungal Divers* 66: 153–165.)

occupational disease, sporotrichosis is frequently observed in farm workers and horticulturists. Animal sporotrichosis has been reported in armadillos, birds, camels, cats, cows, dogs, dolphins, goats, horses, mules, mice, pigs, and rats (Rodrigues et al., 2018). However, cats are particularly susceptible to *S. brasiliensis* and *S. schenckii*. Moreover, human infection can be a result of zoonotic transmission from pet cats (Gremiao et al., 2017; Seyedmousavi et al., 2018). In areas where there are major epizootics in cats (e.g., Brazil), veterinarians and animal trainers are at risk of the disease. There are no records of direct human transfer of infection, but large epizooties (feline horizontal transmission) are frequent in highly endemic areas of Brazil, where *S. brasiliensis* is the prevalent species (Rodrigues et al., 2013b). To a lesser extent, epizooties due to *S. schenckii* have been reported in other countries (Dunstan et al., 1986; Kano et al., 2015).

Sporotrichosis has been reported from all continents, but highly endemic areas are located in temperate, warm, and tropical areas (Zhang et al., 2015). Differences in prevalence are noted and can be mainly explained by the uneven geographical distribution of etiological agents (Chakrabarti et al., 2015). For example, in North America, *S. schenckii* is the main agent in the north-central part of the United States; in Europe, the disease is rare, and most of the cases were reported in France in the early 20th century (Beurmann & Gougerot, 1912). In Asia, India and China concentrate most of the human cases due to *S. globosa* (Zhao et al., 2017), whereas South America, particularly Brazil, has experienced significant emergence of *S. brasiliensis* infection in humans transmitted from cats (Gremiao et al., 2017). Remarkably, epidemics or series of cases have involved all pathogenic species in different endemic areas (Rodrigues et al., 2016b).

In areas where epizootics occur, late diagnosis and treatment of infected cats can lead to rapid transmission of the disease among other animals and humans. On the other hand, socioeconomic factors such as the abandonment of diseased animals on streets and inadequate disposal of dead animal carcasses, in addition to the absence of public health policies to contain the disease, have contributed to sporotrichosis reaching epidemic proportions among humans and animals (Gremiao et al., 2017; Rodrigues et al., 2017). Usually, the increase in the number of cases in felines is followed by a larger number of human cases, which poses a serious public health problem (Rodrigues et al., 2016b).

11.2 Virulence of *Sporothrix* species

The mechanisms involved during the pathogenesis and invasion of *Sporothrix* are still poorly understood and many virulence factors remain unidentified and/or not validated. Among the species that infect mammals, virulence profiles vary according to the characteristics of the agent (Fernandes et al., 2013, 2009a,b) and host immunity (Paixao et al., 2015). *Sporothrix brasiliensis* is by far the most virulent species of the clinical clade, capable of promoting intense tissue invasion and induction of death (Della Terra et al., 2017), whereas *S. schenckii s. str.* has different levels of pathogenicity, ranging from high to low virulence, and *S. globosa* exhibits little or no virulence in murine models (Arrillaga-Moncrieff et al., 2009; Fernandes et al., 2013). Such virulence behavior observed in animal models can be observed in the clinical scenario, and in general, the most severe and atypical clinical outcomes are usually associated with *S. brasiliensis* infection (Almeida-Paes et al., 2014). The environmental *Sporothrix* such as *S. chilensis*, *S. mexicana*, and *S. pallida* have attenuated virulence in murine models with low invasive potential, and the host is able to fight infection within weeks after challenge (Arrillaga-Moncrieff et al., 2009; Rodrigues et al., 2016a).

11.3 Pathogenesis and clinical features

Sporotrichosis is an infection that tends to cause suppuration, fistulization, and ulceration. Usually, a primary lesion, the so-called inoculation canchre, is preceded by a well-defined episode of cutaneous inoculation or trauma, which may be followed by regional nodular lymphangitis.

Fixed cutaneous sporotrichosis is characterized by the presence of a single, localized lesion that frequently appears on the face, neck, trunk, or arms as an ulcer, papule, or verrucous plaque. The initial lesion may have the appearance of an ulcer or a small reddish to purple nodule. Lymphocutaneous sporotrichosis is the classical and most common presentation of the disease, which is characterized by the development of superficial cutaneous lesions that progress along dermal and subcutaneous lymphatic vessels. Lymphocutaneous sporotrichosis and the fixed cutaneous form of the disease together account for more than 95% of the occurrences of sporotrichosis (Estrada-Castanon et al., 2018; Richardson & Warnock, 2012; Rojas et al., 2018). Disseminated cases with or without pulmonary involvement and atypical manifestations with ocular involvement, although not common, are being diagnosed with greater frequency, as shown by reports in the literature (Aung et al. 2013; de Macedo et al. 2015; Orofino-Costa et al. 2013). Usually an impaired immune system due to chronic alcoholism, AIDS, or malignant neoplasia, among others, can predispose an individual to atypical and more severe clinical forms, which can lead to death.

The differential diagnosis of sporotrichosis should consider the diversity of clinical manifestations. In this scenario, in the fixed form, the main differential diagnosis is cutaneous leishmaniasis (de Lima Barros et al., 2005), chromoblastomycosis, histoplasmosis, paracoccidioidomycosis, or tuberculosis, among others (Orofino-Costa et al., 2017). In the lymphocutaneous form, other disorders that present nodular lymphangitis should be investigated, such as atypical mycobacterial infection, especially by *Mycobacterium marinum* (Barros et al., 2011). Currently, the clinical classification adopted is based on the review published by Orofino-Costa et al. (2017) (Table 11.1).

Table 11.1   Clinical Classification of Human Sporotrichosis

Skin	Lymphocutaneous
	Fixed cutaneous
	Multiple inoculation
Mucous membrane	Ocular
	Nasal
	Others
Systemic	Osteoarticular
	Cutaneous disseminated
	Pulmonary
	Neurological
	Other locations/sepsis
Immunoreactive	Erythema nodosum
	Erythema multiforme
	Sweet's syndrome
	Reactive arthritis
Spontaneous regression	

Source: Based on Orofino-Costa, R. C. et al. 2017. *An Bras Dermatol* 92: 606–620.

11.4 Sample collection

Diagnosis of sporotrichosis relies on a combination of clinical presentation, histopathological findings, serology, and culture of the etiologic agents. The main specimens collected for direct fungal microscopy and culture in the lymphocutaneous and fixed cutaneous forms are pus and/ or cutaneous wound secretion. The sample should be collected after skin asepsis with alcohol 70°GL, preferably by aspiration of the pus if there is a floating abscess or accessing the exudate directly from an open wound. Pus or exudate obtained by these methods is immediately plated on culture media. Organic liquids, secretions, and biopsy fragments are stored in a sterile bottle containing sterile saline for culture processing. A biopsy fragment can be surgically collected from the affected skin, mucous membrane, or internal organs, and stored in formalin for processing in the laboratory for histopathology.

11.5 Direct examination: Mycology and histopathology

While most human pathogenic fungi can be identified from potassium hydroxide wet mount slides (KOH 10%–40%), stained smears, or biopsy specimens, such preparations are usually negative in human sporotrichosis due to the paucity of yeast cells (Barros et al., 2011; Richardson & Warnock, 2012). However, when present, these scarce fungal yeast cells can be spherical, oval to elongated "cigar-shaped" (2 to 6 µm), and are positive with periodic acid-Schiff (PAS) staining.

The processed biopsy fragments are usually submitted to impregnation with Grocott-Gomori methenamine silver (GMS) or are stained with PAS (Figure 11.2i–l). Histopathology findings usually reveal a mixed suppurative and granulomatous inflammatory reaction in the dermis and subcutaneous tissue, showing neutrophils, histiocytes, and plasmocytes, often with fibrosis (Barros et al., 2011). On the other hand, feline lesions contain a large fungal load, making direct microscopy through imprint ideal for fast and inexpensive diagnosis of feline sporotrichosis (Miranda et al., 2013; Pereira et al., 2011).

In cats, serial slides of biopsy tissue prepared using PAS or GMS can easily demonstrate the presence of the fungus (Gremião et al., 2015; Miranda et al., 2013). In humans, a peculiar

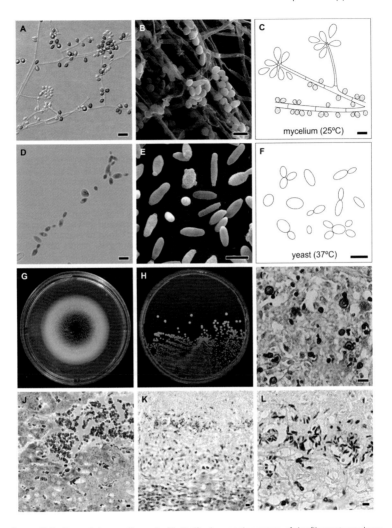

Figure 11.2 *Sporothrix* spp. diagnosis. (A, B, C) microscopic aspects of the filamentous phase of *Sporothrix*, showing hyaline septate hyphae (~ 2 μm in width), hyaline conidia (primary or sympodial conidia) or dark brown conidia (secondary or sessile conidia). (D, E, F) yeast phase of *Sporothrix*, represented by spherical, oval to elongated "cigar-shaped" (2 to 6 μm) yeasts. (G) *Sporothrix* spp. culture at room temperature (25 °C), showing the mycelial form. (H) *Sporothrix* spp. culture at 37 °C, developing the yeast form. (I) *Sporothrix* spp. in tissue stained by PAS (100X), showing spherical yeasts with blastoconidia. (J) *Sporothrix* spp. in tissue stained by GMS (40X). (K, L) *Sporothrix* spp. in tissue showing elongated "cigar-shaped" yeasts (40X and 100X, respectively). Bars = 5 μm.

"asteroid body" has been reported in the pus or section of patients in South America and South Africa, but rarely in the United States and not at all in Europe. "Cigar bodies" are described and are also scarce (Figure 11.2k,l). Such observations are too infrequent to be of diagnostic value (Richardson & Warnock, 2012). Yeast budding cells seen on direct examination, either by KOH or histopathology, can mimic other mycoses caused by dimorphic fungi and certainly are

Table 11.2   Primers Used for Amplification, DNA Sequencing, and Phylogenetic Analysis of *Sporothrix schenckii* and Allied Species

Locus[a]	Primer Name	Primer Sequence (5'à3')	Tm	Primer Ref.	Application			References
					Species ID	Phylogenetic	Genetic Diversity	
CAL	CL1	GARTWCAAGGAGGCCTTCTC	60°C	O'Donnell et al. (2000)	Yes	Yes	Yes	Marimon et al. (2006)
	CL2A	TTTTTGCATCATGAGTTGGAC						
CAL	CAL-Fw	CGCAATGCCAGGCCGAGTCAC	60°C	Rodrigues, et al. (2015e)	Yes	Yes	Yes	Rodrigues et al. (2015e)
	CAL-Rv	ATTTCTGCATCATGAGCTGGAC						
BT2	BT2F	GGYAACCARATHGGTGCYGCY	60°C	Marimon et al. (2006)	Yes	Yes	Yes	Marimon et al. (2006)
	BT2R	ACCCTCRGTGTAGTGACCGGC						
EF1α	EF1-F	CTGAGGCTCGTTACCAGGAG	57°C	Rodrigues et al. (2013b)	Yes	Yes	No	Rodrigues et al. (2013b)
	EF1-R	CGACTTGATGACACCGACAG						
ITS	ITS1	TCCGTAGGTGAACCTTGCGG	52°C	White et al. (1990)	Yes	Yes	No	Zhou et al. (2014)
	ITS4	TCCTCCGCTTATTGATATGC						

[a] *CAL*, calmodulin; *BT2*, β-tubulin; *EF1α*, elongation factor 1-alpha; *ITS*, Internal transcribed spacer.

Table 11.3 Primers Used for Species-Specific Identification of Clinical *Sporothrix* Species DNA

Target Species	Locus	Primer Name	Primer Sequence (5'à3')	Tm	Amplicon
S. brasiliensis	*CAL*	Sbra-F	CCC CCG TTT GAC GCT TGG	TCD[a] (70°C–60°C)	469 bp
		Sbra-R	CCC GGA TAA CCG TGT GTC ATA AT		
S. schenckii	*CAL*	Ssch-F	TTT CGA ATG CGT TCG GCT GG	TCD[a] (70°C–60°C)	331 bp
		Ssch-R	CTC CAG ATC ACC GTG TCA		
S. globosa	*CAL*	Sglo-F	CGC CTA GGC CAG ATC ACC ACT AAG	TCD[a] (70°C–60°C)	243 bp
		Sglo-R	CCA ATG TCT ACC CGT GCT		
S. mexicana	*CAL*	Smex-F	TCT CTG CCG ACA ATT CTT TCT C	TCD[a] (70°C–60°C)	183 bp
		Smex-R	GGA AAG CGG TGG CTA GAT GC		
S. pallida	*CAL*	Spa-F	CGC TGC TTT CCG CCA TTT TCG C	TCD[a] (70°C–60°C)	363 bp
		Spa-R	GCC ATT GTT GTC GCG GTC GAA G		
S. stenoceras	*CAL*	Oste-F	GTG AAC ACC CTC TAT GTA CTT CG	TCD[a] (70°C–60°C)	144 bp
		Oste-R	GTG TAG AGG GGG ATA GAC AGT G		

Source: Adapted from Rodrigues, A. M. et al. 2015a. *PLoS Negl Trop Dis* 9: e0004190.
Abbreviation: CAL, calmodulin.
[a] In the singleplex species-specific PCR, the touchdown protocol (TCD) consisted of two phases: phase 1 included an initial step of 95°C for 5 min, followed by 20 cycles of denaturation at 95°C for 1 min, annealing at variable temperatures for 1 min, and extension at 72°C for 1 min. In the first cycle, the annealing temperature was set to 70°C, and every two cycles the annealing temperature was decreased by 1°C. Phase 2 consisted of 15 cycles at 95°C for 1 min, 60°C for 1 min, and 72°C for 1 min. Finally, the PCR was completed with a final extension at 72°C for 10 minutes.

Aung, A. K., Teh, B. M., McGrath, B. M., Thompson, P. J. 2013. Pulmonary sporotrichosis: Case series and systematic analysis of literature on clinico-radiological patterns and management outcomes. *Med Mycol* 51: 534–544.

Barros, M. B., de Almeida Paes, R., Schubach, A. O. 2011. *Sporothrix schenckii* and sporotrichosis. *Clin Microbiol Rev* 24: 633–654.

Berbee, M. L., Taylor, J. W. 1992. 18S Ribosomal RNA gene sequence characters place the human pathogen *Sporothrix schenckii* in the genus *Ophiostoma*. *Exp Mycol* 16: 87–91.

Bernardes-Engemann, A. R., de Lima Barros, M., Zeitune, T., Russi, D. C., Orofino-Costa, R., Lopes-Bezerra, L. M. 2015. Validation of a serodiagnostic test for sporotrichosis: a follow-up study of patients related to the Rio de Janeiro zoonotic outbreak. *Med Mycol* 53: 28–33.

Beurmann, L., Gougerot, H. 1912. *Les Sporotrichose*. Paris: Librairie Felix Alcan.

Borba-Santos, L. P., Rodrigues, A. M., Gagini, T. B. et al. 2015. Susceptibility of *Sporothrix brasiliensis* isolates to amphotericin B, azoles, and terbinafine. *Med Mycol* 53: 178–188.

Brilhante, R. S., Rodrigues, A. M., Sidrim, J. J. et al. 2016. *In vitro* susceptibility of antifungal drugs against *Sporothrix brasiliensis* recovered from cats with sporotrichosis in Brazil. *Med Mycol* 54: 275–279.

Camacho, E., León-Navarro, I., Rodríguez-Brito, S., Mendoza, M., Niño-Veja, G. A. 2015. Molecular epidemiology of human sporotrichosis in Venezuela reveals high frequency of *Sporothrix globosa*. *BMC Infect Dis* 15: 94.

Carlos, I. Z., Sassa, M. F., da Graca Sgarbi, D. B., Placeres, M. C., Maia, D. C. 2009. Current research on the immune response to experimental sporotrichosis. *Mycopathologia* 168: 1–10.

Chakrabarti, A., Bonifaz, A., Gutierrez-Galhardo, M. C., Mochizuki, T., Li, S. 2015. Global epidemiology of sporotrichosis. *Med Mycol* 53: 3–14.

de Almeida, J. R. F., Jannuzzi, G. P., Kaihami, G. H., Breda, L. C. D., Ferreira, K. S., de Almeida, S. R. 2018. An immunoproteomic approach revealing peptides from *Sporothrix brasiliensis* that induce a cellular immune response in subcutaneous sporotrichosis. *Sci Rep* 8: 4192.

de Beer, Z. W., Duong, T. A., Wingfield, M. J. 2016. The divorce of *Sporothrix* and *Ophiostoma*: Solution to a problematic relationship. *Stud Mycol* 83: 165–191.

de Beer, Z. W., Harrington, T. C., Vismer, H. F., Wingfield, B. D., Wingfield, M. J. 2003. Phylogeny of the *Ophiostoma stenoceras–Sporothrix schenckii* complex. *Mycologia* 95: 434–441.

de Lima Barros, M. B., Schubach, A., Francesconi-do-Valle, A. C. et al. 2005. Positive Montenegro skin test among patients with sporotrichosis in Rio De Janeiro. *Acta Tropica* 93: 41–47.

de Lima Barros, M. B., Schubach, A. O., de Oliveira, R. V. C., Martins, E. B., Teixeira, J. L., Wanke, B. 2011. Treatment of cutaneous sporotrichosis with Itraconazole—Study of 645 patients. *Clin Infect Dis* 52: e200–e06.

Della Terra, P. P., Rodrigues, A. M., Fernandes, G. F., Nishikaku, A. S., Burger, E., de Camargo, Z. P. 2017. Exploring virulence and immunogenicity in the emerging pathogen *Sporothrix brasiliensis*. *PLoS Negl Trop Dis* 11: e0005903.

de Macedo, P. M., Sztajnbok, D. C., Camargo, Z. P. et al. 2015. Dacryocystitis due to *Sporothrix brasiliensis*: A case report of a successful clinical and serological outcome with low-dose potassium iodide treatment and oculoplastic surgery. *Br J Dermatol* 172: 1116–1119.

de Meyer, E. M., de Beer, Z. W., Summerbell, R. C. et al. 2008. Taxonomy and phylogeny of new wood- and soil-inhabiting *Sporothrix* species in the *Ophiostoma stenoceras-Sporothrix schenckii* complex. *Mycologia* 100: 647–661.

de Oliveira, M. M., Sampaio, P., Almeida-Paes, R., Pais, C., Gutierrez-Galhardo, M. C., Zancope-Oliveira, R. M. 2012. Rapid identification of *Sporothrix* species by T3B fingerprinting. *J Clin Microbiol* 50: 2159–2162.

Dixon, D M., Salkin, I. F., Duncan, R. A. et al. 1991. Isolation and characterization of *Sporothrix schenckii* from clinical and environmental sources associated with the largest U.S. epidemic of sporotrichosis. *J Clin Microbiol* 29: 1106–1113.

Dunstan, R. W., Langham, R. F., Reimann, K. A., Wakenell, P. S. 1986. Feline sporotrichosis: A report of five cases with transmission to humans. *J Am Acad Dermatol* 15: 37–45.

Espinel-Ingroff, A., Abreu, D. P. B., Almeida-Paes, R. et al. 2017. Multicenter and international study of MIC/MEC distributions for definition of epidemiological cutoff values (ECVs) for species of *Sporothrix* identified by molecular methods. *Antimicrob Agents Chemother* 61: e01057–e01017.

Estrada-Castanon, R., Chavez-Lopez, G., Estrada-Chavez, G., Bonifaz, A. 2018. Report of 73 cases of cutaneous sporotrichosis in Mexico. *An Bras Dermatol* 93: 907–909.

Fernandes, G. F., Do Amaral, C. C., Sasaki, A., Godoy, P. M., De Camargo, Z. P. 2009a. Heterogeneity of proteins expressed by Brazilian *Sporothrix schenckii* isolates. *Med Mycol* 47: 855–861.

Fernandes, G. F., dos Santos, P. O., do Amaral, C. C., Sasaki, A. A., Godoy-Martinez, P., Camargo, Z. P. 2009b. Characteristics of 151 Brazilian *Sporothrix schenckii* isolates from 5 different geographic regions of brazil: A forgotten and re-emergent pathogen. *Open Mycol J* 3: 48–58.

Fernandes, G. F., dos Santos, P. O., Rodrigues, A. M., Sasaki, A. A., Burger, E., de Camargo, Z. P. 2013. Characterization of virulence profile, protein secretion and immunogenicity of different *Sporothrix schenckii* sensu stricto isolates compared with *S. globosa* and *S. brasiliensis* species. *Virulence* 4: 241–249.

Fernandes, G. F., Lopes-Bezerra, L. M., Bernardes-Engemann, A. R. et al. 2011. Serodiagnosis of sporotrichosis infection in cats by enzyme-linked immunosorbent assay using a specific antigen, SsCBF, and crude exoantigens. *Vet Microbiol* 147: 445–549.

Gremião, I. D., Menezes, R. C., Schubach, T. M., Figueiredo, A. B., Cavalcanti, M. C., Pereira, S. A. 2015. Feline sporotrichosis: Epidemiological and clinical aspects. *Med Mycol* 53: 15–21.

Gremiao, I. D., Miranda, L. H., Reis, E. G., Rodrigues, A. M., Pereira, S. A. 2017. Zoonotic epidemic of sporotrichosis: Cat to human transmission. *PLoS Pathog* 13: e1006077.

Hu, S., Chung, W. H., Hung, S. I. et al. 2003. Detection of *Sporothrix schenckii* in clinical samples by a nested PCR assay. *J Clin Microbiol* 41: 1414–1418.

Kano, R., Okubo, M., Siew, H. H., Kamata, H., Hasegawa, A. 2015. Molecular typing of *Sporothrix schenckii* isolates from cats in Malaysia. *Mycoses* 58: 220–224.

Kauffman, C. A., Bustamante, B., Chapman, S. W., Pappas, P. G. 2007. Clinical practice guidelines for the management of sporotrichosis: 2007 update by the Infectious Diseases Society of America. *Clin Infect Dis* 45: 1255–1265.

Kwon-Chung, J. K., Bennett, J. E. 1992. *Medical Mycology*. Philadelphia: Lea & Febiger.

Marimon, R., Cano, J., Gené, J., Sutton, D. A., Kawasaki, M., Guarro, J. 2007. *Sporothrix brasiliensis*, *S. globosa*, and *S. mexicana*, three new *Sporothrix* species of clinical interest. *J Clin Microbiol* 45: 3198–3206.

Marimon, R., Gené, J., Cano, J., Trilles, L., Lazéra, M. S., Guarro, J. 2006. Molecular phylogeny of *Sporothrix schenckii*. *J Clin Microbiol* 44: 3251–3256.

Mayorga, R., Caceres, A., Toriello, C. et al. 1978. An endemic area of sporotrichosis in Guatemala. *Sabouraudia* 16: 185–198.

Miranda, L. H., Conceicao-Silva, F., Quintella, L. P., Kuraiem, B. P., Pereira, S. A., Schubach, T. M. 2013. Feline sporotrichosis: Histopathological profile of cutaneous lesions and their correlation with clinical presentation. *Comp Immunol Microbiol Infect Dis* 36: 425–432.

Montagnoli, C., Bozza, S., Bacci, A. et al. 2003. A role for antibodies in the generation of memory antifungal immunity. *Eur J Immunol* 33: 1193–1204.

Morrison, A. S., Lockhart, S. R., Bromley, J. G., Kim, J. Y., Burd, E. M. 2013. An environmental *Sporothrix* as a cause of corneal ulcer. *Med Mycol Case Rep* 2: 88–90.

Moussa, T. A., Kadasa, N. M., Al Zahrani, H. S. et al. 2017. Origin and distribution of *Sporothrix globosa* causing sapronoses in Asia. *J Med Microbiol* 66: 560–569.

Nascimento, R. C., Almeida, S. R. 2005. Humoral immune response against soluble and fractionate antigens in experimental sporotrichosis. *FEMS Immunol Med Microbiol* 43: 241–247.

Nascimento, R. C., Espíndola, N. M., Castro, R. A. et al. 2008. Passive immunization with monoclonal antibody against a 70-kDa putative adhesin of *Sporothrix schenckii* induces protection in murine sporotrichosis. *Eur J Immunol* 38: 3080–3089.

O'Donnell, K., Nirenberg, H., Aoki, T., Cigelnik, E. 2000. A multigene phylogeny of the *Gibberella fujikuroi* species complex: Detection of additional phylogenetically distinct species. *Mycoscience* 41: 61–78.

Oliveira, M. M., Almeida-Paes, R., Muniz, M. M., Gutierrez-Galhardo, M. C., Zancope-Oliveira, R. M. 2011. Phenotypic and molecular identification of *Sporothrix* isolates from an epidemic area of sporotrichosis in Brazil. *Mycopathologia* 172: 257–267.

Oliveira, M. M., Santos, C., Sampaio, P. et al. 2015. Development and optimization of a new MALDI-TOF protocol for identification of the *Sporothrix* species complex. *Res Microbiol* 166: 102–110.

Orofino-Costa, R., de Macedo, P. M., Bernardes-Engemann, A. R. 2015. Hyperendemia of sporotrichosis in the Brazilian Southeast: Learning from clinics and therapeutics. *Curr Fungal Infect Rep* 9: 220–228.

Orofino-Costa, R., Unterstell, N., Carlos Gripp, A. et al. 2013. Pulmonary cavitation and skin lesions mimicking tuberculosis in a HIV negative patient caused by *Sporothrix brasiliensis*. *Med Mycol Case Rep* 2: 65–71.

Orofino-Costa, R. C., Macedo, P. M., Rodrigues, A. M., Bernardes-Engemann, A. R. 2017. Sporotrichosis: An update on epidemiology, etiopathogenesis, laboratory and clinical therapeutics. *An Bras Dermatol* 92: 606–620.

Paixao, A. G., Galhardo, M. C., Almeida-Paes, R. et al. 2015. The difficult management of disseminated *Sporothrix brasiliensis* in a patient with advanced AIDS. *AIDS Res Ther* 12: 16.

Pereira, S. A., Menezes, R. C., Gremião, I. D. et al. 2011. Sensitivity of cytopathological examination in the diagnosis of feline sporotrichosis. *J Feline Med Surg* 13: 220–223.

Pereira, S. A., Passos, S. R., Silva, J. N. et al. 2010. Response to azolic antifungal agents for treating feline sporotrichosis. *Vet Rec* 166: 290–294.

Portuondo, D. L., Batista-Duharte, A., Ferreira, L. S. et al. 2016. A cell wall protein-based vaccine candidate induce protective immune response against *Sporothrix schenckii* infection. *Immunobiology* 221: 300–309.

Ramírez-Soto, M., Aguilar-Ancori, E., Tirado-Sánchez, A., Bonifaz, A. 2018. Ecological determinants of sporotrichosis etiological agents. *J Fungi* 4: E95.

Reis, E. G., Schubach, T. M., Pereira, S. A. et al. 2016. Association of itraconazole and potassium iodide in the treatment of feline sporotrichosis: A prospective study. *Med Mycol* 54: 684–690.

Richardson, M. D., Warnock, D. W. 2012. *Fungal Infection: Diagnosis and Management.* 4th ed. Chichester: Wiley-Blackwell.

Rodrigues, A. M., Bagagli, E., de Camargo, Z. P., Bosco, M. S. G. 2014a. *Sporothrix schenckii sensu stricto* isolated from soil in an armadillo's burrow. *Mycopathologia* 177: 199–206.

Rodrigues, A. M., Choappa, R. C., Fernandes, G. F., De Hoog, G. S., Camargo, G. S., 2016a. *Sporothrix chilensis* sp. nov. (Ascomycota: Ophiostomatales), a soil-borne agent of human sporotrichosis with mild-pathogenic potential to mammals. *Fungal Biol* 120: 246–264.

Rodrigues, A. M., de Hoog, G. S., Camargo, Z. P. 2014b. Genotyping species of the *Sporothrix schenckii* complex by PCR-RFLP of calmodulin. *Diagn Microbiol Infect Dis* 78: 383–387.

Rodrigues, A. M., de Hoog, G. S., Camargo, Z. P. 2018. Feline Sporotrichosis. In *Emerging and Epizootic Fungal Infections in Animals*, edited by Seyedmousavi, S. de Hoog, G. S., Guillot, J., Verweij, P. E. 199–231, Cham: Springer International Publishing.

Rodrigues, A. M., de Hoog, G. S., de Camargo, Z. P. 2015a. Molecular diagnosis of pathogenic *Sporothrix* species. *PLoS Negl Trop Dis* 9: e0004190.

Rodrigues, A. M., de Hoog, G. S., de Camargo, Z. P. 2016b. *Sporothrix* species causing outbreaks in animals and humans driven by animal-animal transmission. *PLoS Pathog* 12: e1005638.

Rodrigues, A. M., de Hoog, G. S., Pires, D. C. et al. 2014c. Genetic diversity and antifungal susceptibility profiles in causative agents of sporotrichosis. *BMC Infect Dis* 14: 219.

Rodrigues, A. M., de Hoog, G. S., Zhang, Y., Camargo, Z. P. 2014d. Emerging sporotrichosis is driven by clonal and recombinant *Sporothrix* species. *Emerg Microbes Infect* 3: e32.

Rodrigues, A. M., de Hoog, S., de Camargo, Z. P. 2013a. Emergence of pathogenicity in the *Sporothrix schenckii* complex. *Med Mycol* 51: 405–412.

Rodrigues, A. M., de Melo Teixeira, M., de Hoog, G. S., Schubach, T. M., Pereira, S. A., Fernandes, G. F. 2013b. Phylogenetic analysis reveals a high prevalence of *Sporothrix brasiliensis* in feline sporotrichosis outbreaks. *PLoS Negl Trop Dis* 7: e2281.

Rodrigues, A. M., Fernandes, G. F., Araujo, L. M. et al. 2015b. Proteomics-based characterization of the humoral immune response in sporotrichosis: Toward discovery of potential diagnostic and vaccine antigens. *PLoS Negl Trop Dis* 9: e0004016.

Rodrigues, A. M., Fernandes, G. F., de Camargo, Z. P. 2017. Sporotrichosis. In *Emerging and Re-Emerging Infectious Diseases of Livestock*, edited by Bayry, J. 391–421, Cham: Springer International Publishing.

Rodrigues, A. M., Kubitschek-Barreira, P. H., Fernandes, G. F., Almeida, S. R., Lopes-Bezerra, L. M., Camargo, Z. P. 2015c. Two-dimensional gel electrophoresis data for proteomic profiling of *Sporothrix* yeast cells. *Data in Brief* 2: 32–38.

Rodrigues, A. M., Kubitschek-Barreira, P. H., Fernandes, G. F., de Almeida, S. R., Lopes-Bezerra, L. M., de Camargo, Z. P. 2015d. Immunoproteomic analysis reveals a convergent humoral response signature in the *Sporothrix schenckii* complex. *J Proteomics* 115: 8–22.

Rodrigues, A. M., Najafzadeh, M. J., de Hoog, G. S., de Camargo, Z. P. 2015e. Rapid identification of emerging human-pathogenic *Sporothrix* species with rolling circle amplification. *Front Microbiol* 6: 1385.

Rodriguez-Brito, S., Camacho, E., Mendoza, M., Nino-Veja, G. A. 2015. Differential identification of *Sporothrix* spp. and *Leishmania* spp. by conventional PCR and qPCR in multiplex format. *Med Mycol* 53: 22–27.

Rojas, O. C., Bonifaz, A., Campos, C., Trevino-Rangel, R. J., Gonzalez-Alvarez, R., Gonzalez, G. M. 2018. Molecular identification, antifungal susceptibility, and geographic origin of clinical strains of *Sporothrix schenckii* complex in Mexico. *J Fungi* 4: 3.

Romani, L. 2011. Immunity to fungal infections. *Nat Rev Immunol* 11: 275–288.

Ruiz-Baca, E., Hernandez-Mendoza, G., Cuellar-Cruz, M., Toriello, C., Lopez-Romero E., Gutierrez-Sanchez, G. 2014. Detection of 2 immunoreactive antigens in the cell wall of *Sporothrix brasiliensis* and *Sporothrix globosa*. *Diagn Microbiol Infect Dis* 79: 328–330.

Ruiz-Baca, E., Mora-Montes, H. M., Lopez-Romero, E., Toriello, C., Mojica-Marin, V., Urtiz-Estrada, N. 2011. 2D-immunoblotting analysis of *Sporothrix schenckii* cell wall. *Mem Inst Oswaldo Cruz* 106: 248–250.

Schubach, T. M., Schubach, A., Okamoto, T., Barros, M. B., Figueiredo, F. B., Cuzzi, T. 2004. Evaluation of an epidemic of sporotrichosis in cats: 347 cases (1998–2001). *J Am Vet Med Assoc* 224: 1623–1629.

Scott, E. N., Muchmore, H. G. 1989. Immunoblot analysis of antibody responses to *Sporothrix schenckii*. *J Clin Microbiol* 27: 300–304.

Seyedmousavi, S., Bosco, S. M. G., de Hoog, S. et al. 2018. Fungal infections in animals: A patchwork of different situations. *Med Mycol* 56: S165–S87.

Silva-Vergara, M. L., Camargo, Z. P., Silva, P. F. et al. 2012. Disseminated *Sporothrix brasiliensis* infection with endocardial and ocular involvement in an HIV-infected patient. *Am J Trop Med Hyg* 86: 477–480.

Singhal, N., Kumar, M., Kanaujia, P. K., Virdi, J. S. 2015. MALDI-TOF mass spectrometry: An emerging technology for microbial identification and diagnosis. *Front Microbiol* 6: 791–791.

Teixeira, P. A., de Castro, R. A., Nascimento, R. C., Tronchin, G., Torres, A. P., Lazéra, M. 2009. Cell surface expression of adhesins for fibronectin correlates with virulence in *Sporothrix schenckii*. *Microbiology* 155: 3730–3738.

White, T.J., Bruns, T., Lee, S., Taylor, J. 1990. Amplification and direct sequencing of fungal ribosomal RNA genes for phylogenetics. In *PCR Protocols: A Guide to Methods and Applications*, Innis, M. A., Gelfand, D. H., Shinsky, J. J., White, T. J. (eds), New York: Academic Press.

Zhang, Y., Hagen, F., Stielow, B. et al. 2015. Phylogeography and evolutionary patterns in *Sporothrix* spanning more than 14,000 human and animal case reports. *Persoonia* 35: 1–20.

Zhao, L., Cui, Y., Zhen, Y. et al. 2017. Genetic variation of *Sporothrix globosa* isolates from diverse geographic and clinical origins in China. *Emerg Microbes Infect* 6: e88.

Zhou, X., Rodrigues, A. M., Feng, P., De Hoog, G. S. 2014. Global ITS diversity in the *Sporothrix schenckii* complex. *Fungal Divers* 66: 153–165.

12
Histoplasma capsulatum

Rosely Maria Zancopé-Oliveira, Claudia Vera Pizzini,
Marcos de Abreu Almeida, and Rodrigo de Almeida Paes

Contents

12.1 Introduction

Histoplasmosis, the disease caused by *H. capsulatum*, is arguably the most common fungal respiratory infection worldwide, and is endemic in the Americas. The disease burden of pulmonary histoplasmosis is among the highest of any disease caused by a primary fungal pathogen (Wheat et al., 2016). In 5%–10% of cases, the infection progresses in the lung or disseminates to visceral organs and can be difficult to treat with antifungal drugs. Patients who appear to recover may suffer a recurrence months to years later, particularly if they are immunocompromised, mainly individuals with AIDS (Nacher et al., 2016).

12.2 Collection, transport, and processing of clinical samples

Diagnosis of histoplasmosis is a challenge and requires a multiprofessional approach. Laboratory, radiologic, histopathologic, microbiologic, serologic, and molecular analyses are necessary for a reliable diagnosis. The fungus can be isolated from sputum, bronchoalveolar lavage (BAL), bone marrow, blood, urine, and biopsied tissues, depending on the clinical manifestation of the disease. For serology, serum, plasma, cerebrospinal fluid (CSF), and urine can be used for antigen and antibody detection (Azar & Hage, 2017). The following guidelines should be considered in the collection of these samples (Moraes et al., 2009):

- *Sputum*: The patient must perform oral hygiene before the collection. Both induced and spontaneous sputum are acceptable. Sample must be collected in a sterile universal container and sent to the laboratory as soon as possible.
- *BAL*: Collection must be performed by a physician, following bronchoscopy guidelines. Sample must be collected in a sterile vial and sent to the laboratory as soon as possible.
- *Bone marrow*: Collection must be performed by a physician, following the specific guidelines. Sample must be collected in a sterile tube with heparin and sent to the laboratory as soon as possible.
- *Blood*: Collection must be performed from a peripheral venipuncture, according to the specific guidelines. For whole blood culture proceedings, sample must be collected in blood

culture bottles, and for buffy coat culture, sample must be collected in EDTA tubes. For serology tests, serum or plasma should be collected after blood centrifugation. Samples for microbiologic analyses should be sent to the laboratory as soon as possible.

- *Urine*: The patient must perform local hygiene before the collection. Preferably, early morning samples should submitted for culture, but for antigen detection, samples collected at other times are acceptable as well. Midstream urine should be collected by the clean-catch method and put in a sterile universal container. Samples from urinals or bedpans; bags of catheterized patients; leaky containers; unlabeled specimens; and unrefrigerated, unpreserved samples over 2 hours old must be rejected for microbiologic evaluation.
- *Biopsies*: Collection must be performed by a physician, following the specific medical procedures according to the biopsy tissue. For culture procedures, sample must be collected in a vial with sterile physiologic saline solution, and for histopathology in a vial with formaldehyde 10%. Samples for microbiologic analyses should be sent to the laboratory as soon as possible.
- *CSF*: Collection must be performed by a physician, under specific recommendations. Sample must be collected in a sterile vial and sent to the laboratory as soon as possible. In the laboratory, sputum, BAL, and urine samples must be centrifuged, the supernatant discarded, and the pellet used for direct examination and culture. Tissue fragments must be minced in small fragments with a surgical scissors. Blood culture bottles should be incubated in automated systems for at least 30 days. Buffy coat must be collected from the second layer after blood centrifugation and discard of the superior supernatant (plasma). Centrifugation is not required for urine and CSF samples, when indicated for serologic analyses.

12.3 Direct examination and culture

Direct examination with 10% KOH has very low sensitivity in the diagnosis of histoplasmosis. However, it should be performed for a differential diagnosis with other mycotic infections, such as paracoccidioidomycosis (Zancopé-Oliveira et al., 2014). The sensitivity of this method can be increased when specimens such as bronchial aspirates, bone marrow, biopsy, or peripheral blood smear (notably the buffy coat) are staining with Giemsa, May-Grünwald Giemsa (MGG), Wright, periodic acid of Schiff (PAS), or Gomori methenamine silver. *Histoplasma capsulatum* appears as intra- and extracellular tiny round or oval bodies from 1 to 4 μm in diameter, with a recognizable clear halo surrounding a central or eccentric stained chromatin (Couppié et al., 2006). Calcofluor white is useful for *Histoplasma capsulatum* detection in clinical samples, but since this fluorescent stain binds nonspecifically to fungal chitins, visualization of yeasts in this test is not pathognomonic (Azar & Hage, 2017).

The gold-standard diagnostic method in histoplasmosis is fungal isolation and identification in culture. Nevertheless, this method has several limitations, such as a prolonged incubation time for fungal growth, which usually ranges between 2 and 3 weeks, but may take up to 6 weeks, and sensitivity is dependent on the nature of the clinical sample, the clinical manifestation of histoplasmosis, the immune status of the patient, and the fungal burden in the collected sample (Azar & Hage, 2017; Guimarães et al., 2006). Positive clinical samples incubated in Sabouraud dextrose agar or mycobiotic/mycosel agar at 25°C–30°C yield mycelial colonies composed by hyaline, septate hyphae presenting smooth globose microconidia 2–5 μm in diameter, and tuberculate macroconidia 7–15 μm in diameter. Since other fungi can also produce tuberculate macroconidia, conversion of *H. capsulatum* to its yeast form is required. Incubation of the mold form at 37°C in enriched culture media such as MLGema (Fressatti et al., 1992) will result in the appearance of yeast-like colonies composed of hyaline small round narrow-budding yeasts of 2–4 μm in diameter. Figure 12.1 illustrates *H. capsulatum* in both morphologies. Although necessary, the laboratory confirmation of the dimorphic nature of *H. capsulatum* is time-consuming and presents a low rate of success. To improve the identification of *H. capsulatum* cultures, exoantigen tests, molecular probes, and a MALDI-TOF reference database are available (Azar & Hage, 2017; Valero et al., 2017).

12.4 Cytopathology and histopathology

Fine-needle aspiration is a safe diagnostic technique that can yield a cytodiagnosis of histoplasmosis. In fact, examination of aspirates and fluids rather than tissue with preserved

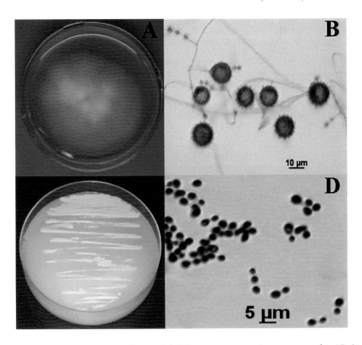

Figure 12.1   *Histoplasma capsulatum*. (A) Culture on potato dextrose agar for 15 days at 25°C. (B) Mycelial colonies composed by hyaline, septate hyphae presenting smooth globose microconidia and tuberculate macroconidia, 40×. (C) Culture on MLGema media at 37°C for 15 days. (D) Yeast-like colonies composed of hyaline small round narrow-budding yeasts, 100×.

architecture can provide presumptive evidence for histoplasmosis. When stained with GMS or PAS, the cytological preparation will often show narrow-based budding yeast cells, mainly within macrophages. In tissues, *H. capsulatum* yeast cells are mainly found phagocytosed within macrophages or histiocytes, often forming clusters, but may sometimes be seen in extracellular spaces. The histopathology examination usually shows incomplete granulomas and/or fibrosis rather than a well-formed pyogranulomatous reaction. In these cases, negative cultures, the absence of histoplasmosis symptoms, and *H. capsulatum* antigenemia can help to distinguish between resolved, previous disease and active infection (Azar & Hage, 2017).

12.5 Immunodiagnostic tests

In the absence of mycological methods, serology is an important tool for the presumptive diagnosis of histoplasmosis, since it indirectly evaluates the existence of the pathogen in the host through the detection of antibodies and/or antigens, as well having a rapid turnaround time, providing information indicative not only for diagnosis but also for monitoring the patient's course (Scheel & Gómez, 2014). There are many advantages to the use of serology for diagnosis of invasive fungal infection, including histoplasmosis. First, results may be positive when cultures are negative or clinical samples are difficult to obtain. Second, if positive, results of serology may reduce the requirement for culture of potentially hazardous fungi, and, finally, serology requires non-invasive clinical specimens. Sensitivity and specificity of some serologic tests could be a disadvantage of serology. A negative serologic test should not exclude the presence of fungal infection; false positives may occur with some tests in the setting of others endemic fungal infections. The sensitivity is dependent on the type of disease and the timing of testing relative to the disease process (Kozel & Wickes, 2014).

12.5.1 Antibody detection

Detection of fungal-specific antibodies can provide evidence of previous exposure to endemic fungi, which can help identify the etiologic agent of an ongoing disease (Powers-Fletcher & Hanson 2016). Anti-*Histoplasma* antibodies require 4–8 weeks to be measurable in peripheral blood (Wheat et al., 2016) and may be negative in immunocompromised patients, especially those who have undergone solid organ transplantation (Hage et al., 2011). Antibody testing is most useful for subacute, chronic, and mediastinal forms of histoplasmosis, in which circulating antibodies are present and the sensitivity of antigen detection is suboptimal (Azar & Hage, 2017). However, it is also useful for disseminated forms of histoplasmosis depending on the applied test for diagnosis (Almeida, 2015). Serology for histoplasmosis can be useful even in endemic areas for this disease, where less than 5% of individuals have positive serology on routine tests (Wheat et al., 1982).

Although several serologic tests have been developed for the diagnosis of histoplasmosis, such as the latex agglutination test (LA-Histo antibody system, Immuno-Mycologics, Inc.) and several enzyme immunoassays in different formats, their sensitivity ranges from 64% to 100%, but they present high levels of cross-reactivity, making interpretation difficult (Guimarães et al., 2006). Therefore, the immunodiffusion (ID), complement fixation (CF), and enzyme immunoassay (EIA) tests and Western blot (WB) are considered the main techniques for antibody detection used in laboratory routine due to their convenience, availability, and accuracy. The ID test detects anti-M and H antigens of *H. capsulatum*–forming precipitins (M, H, or both bands). The sensitivity of ID varies between 70% and 95%, depending on the clinical form, and specificity is 100% (Scheel & Gómez, 2014). The M band is detectable in most patients with acute histoplasmosis (80%) and in some of the individuals sensitized after the cutaneous test with histoplasmin, not being able to distinguish active disease. The H band is found in less than 20% of patients, and when present confirms acute infection (Azar & Hage, 2017). The presence of both the M and H precipitins in a serum sample is considered conclusive for the diagnosis of this mycosis.

The CF method is a quantitative test measuring antibodies to mycelial (histoplasmin) and yeast antigens. The sensitivity of CF is variable, 72%–95%, depending on the antigen phase used. The specificity varies between 70% and 80%, and cross-reactions may occur with other fungal infections such as blastomycosis, candidosis, and paracoccidioidomycosis (Scheel & Gómez, 2014). Titers of 1:8 are considered a positive reaction, indicating previous exposure to *H. capsulatum*. Titers of 1:32 or a fourfold or greater increase in antibody titers between acute and convalescent sera are indicative of active infection. Titers generally decrease with disease resolution, but the decay is slow and often incomplete, making antibody detection impractical as a tool to evaluate treatment response (Azar & Hage, 2017).

Antibodies to *H. capsulatum* can also be detected by immunoenzymatic assay such as Western blot. Western blot technology using the purified and deglycosylated histoplasmin antigen demonstrates a significantly improvement when compared with the native glycosylated antigen (Pizzini et al., 1999), increasing the sensitivity from 45% to 90% in serum samples from acute form of histoplasmosis. This method using the purified and deglycosylated histoplasmin antigen has been further validated, presenting 94.9% and 94.1% of sensitivity and specificity, respectively. Therefore, this method can be considered a useful tool in the diagnosis of histoplasmosis even in situations where laboratory facilities are relatively limited, since WB-sensitized strips may be stored at room temperature up to 5 years, presenting reactivity without compromising the accuracy of the test (Almeida et al., 2016). Moreover, WB is faster than ID and CF used in diagnostic routine, and should be applied in microbiology laboratories since it has almost perfect reproducibility, producing repeated and consistent results.

12.5.2 Antigen detection

Antigen detection tests may be more effective than antibody testing for diagnosing histoplasmosis. During *H. capsulatum* infection, antigens can be released by the fungal cells and detected in body fluids such as serum (blood); pleural, bronchoalveolar, and cerebrospinal fluid; and urine. In addition, their detection has become a leading modality to diagnose histoplasmosis (Azar & Hage, 2017), mainly in countries where antigen tests are available. Detection of antigen may be more effective and useful in acute disease as well as in patients with impaired immunity. In general, patients with disseminated histoplasmosis have high levels of antigenuria, and it is useful for diagnosis and for monitoring the response to therapy. A decreasing level of antigen titers is directly correlated with improvement of the patient's clinical condition. Studies have shown antigen detection in urine has generally proven to be slightly more sensitive than in serum samples across all manifestations of histoplasmosis. Combining both urine and serum testing increases the likelihood of antigen detection (Azar & Hage, 2017; Scheel & Gómez, 2014).

The first described tests for antigen detection used fluorescent antibodies for the screening and identification of *H. capsulatum* in sputum (Lynch & Plexico, 1962). A radioimmunoassay (RIA) for the detection of *H. capsulatum* antigen based on the detection of a polysaccharide antigen from *H. capsulatum* (HPA) in urine and serum specimens was posteriorly developed (Wheat et al., 1986), showing high effectiveness, especially in patients with disseminated histoplasmosis/AIDS. The improvement of this method has resulted in increased levels of detection of HPA, mainly in urine (95%) sera (86%) of AIDS/disseminated histoplasmosis patients (Guimarães et al., 2006).

Several EIA protocols have been described for antigen detection based on the detection of the *H. capsulatum* polysaccharide antigen on various biological specimens as well as in different clinical forms of histoplasmosis. Among them, the MiraVista EIA *Histoplasma* antigen test (MiraVista Diagnostics, Indianapolis, IN) provides a high level of sensitivity in disseminated histoplasmosis (91.8%), followed by chronic and acute pulmonary histoplasmosis (85.5% and 83%, respectively). Sensitivity for subacute histoplasmosis was the lowest (30.3%) (Hage et al., 2011). However, this test requires that all specimens be submitted to MiraVista Diagnostics for testing, which can delay result reporting, consequently affecting patient management (Theel et al., 2015), and it is unavailable to most areas of the world where this fungus is endemic (Scheel & Gómez, 2014). Posteriorly, the IMMY ALPHA ELISA kit (IMMY, Norman, OK), a two-step sandwich-type immunoenzymatic assay using polyclonal antibody that quantitatively detects *Histoplasma* antigens in urine samples, was developed for use at local facilities, but presented sensitivity and specificity lower than those of the MiraVista assay (Zhang et al., 2013). However, a subsequent improvement on the analyte-specific reagent (ASR) *H. capsulatum* antigen EIA (IMMY) (Azar & Hage, 2017) has increased the sensitivity and specificity of this test, as well as agreement with the MiraVista EIA (Theel et al., 2015).

Similar EIA assays to detect *H. capsulatum* antigenuria in immunocompromised patients have been described, but are not commercially available (Scheel & Gómez, 2014). The most well evaluated was an antigen-capture EIA for the detection of circulating *H. capsulatum* antigen in the urine of patients with histoplasmosis and AIDS co-infection and its correlation with clinical improvement during the therapeutic follow-up. The test shows 86.0% sensitivity and 94.0% specificity, with considerable potential to monitor and evaluate the antifungal therapy response in patients with AIDS and disseminated histoplasmosis, especially those with severe immunodeficiency (Caceres et al., 2014).

Nowadays, a lateral flow test for *Histoplasma* antigen detection is also under development by IMMY, and in the future, it will be of possible use as point-of-care diagnosis for histoplasmosis, as it is available for cryptococcosis.

12.6 Molecular diagnostics

In order to circumvent the limitations of histoplasmosis diagnosis, molecular biology techniques have been contributing to the development of approaches that indirectly detect *H. capsulatum* in clinical specimens by identifying specific nucleic acids, offering greater rapidity as compared to other diagnostics as well as high sensitivity and specificity, especially for AIDS patients.

Molecular assays to detect *H. capsulatum* in human specimens can be pan-fungal or organism specific. Pan-fungal assays utilizing the non-translated ribosomal internal transcribed spacer (ITS) gene regions of rDNA are applied to discriminate fungal species, being recommended as the universal DNA barcode marker for fungi (Irinyi et al., 2015). Several PCR-based methods have been described for *H. capsulatum* using different targets, including conventional, nested, and real-time PCR (Falci et al., 2017). Table 12.1 summarizes the results of published molecular assays and shows that the sensitivity ranges from 33% to 100% and the specificity from 96% to 100%. It also shows that the most frequently used molecular target for specific detection of *H. capsulatum* is a locus that encodes a 100 kDa protein, Hcp100. Indeed, molecular methods are more sensitive and specific than immunological or serological tests. However, they are not commercially available, limiting the generalization of *Histoplasma* PCR results (Azar & Hage, 2017; Falci et al., 2017).

Table 12.1 Polymerase Chain Reaction (PCR) Assays for *H. capsulatum* Detection in Clinical Samples

Method	Gene Target	Source	Samples (n)	Sensitivity (%)	Specificity (%)	Reference
Nested PCR	Hcp 100	FFPE	62 (29 *Histoplasma*-positive biopsies)	69	100	Bialek et al. (2002)
Semi-nested PCR	H antigen	Blood and mucosae	30 (6 culture-positive samples)	NR	NR	Bracca et al. (2003)
Real-time PCR	ITS1 of rDNA	Sera	10 (4 culture-positive samples)	70	100	Buitrago et al. (2006)
PCR-EIA	Hcp 100	Urine	76 (51 *Histoplasma*-positive antigen)	80	100	Tang et al. (2006)
Nested PCR	Hcp 100	Several	40 (15 culture-positive samples)	100	100	Maubon et al. (2007)
Nested PCR	Hcp 100	Blood	31 (19 culture-positive samples)	89	96	Toranzo et al. (2007)
Nested PCR	Hcp 100	Several	146 (67 culture-positive samples)	100	92–95	Muñoz et al. (2010)
Real-time PCR	ITS of rDNA	Several	348 (71 culture-positive samples)	96	96	Simon et al. (2010)
Real-time PCR	GAPDH	Several	797 (15 culture-positive samples)	73	100	Babady et al. (2011)
PCR	M antigen	Several	7 (1 culture-positive sample, and 3 *Histoplasma*-positive serology)	NR	NR	Ohno et al. (2013)
PCR	18S, 5.8S rDNA, Hcp 100	Serum and blood	40 (14 culture-positive samples)	70–98	100	Dantas et al. (2013)
PCR and FISH	rRNA	Blood	33 (3 culture-positive samples)	100	100	Silva et al. (2015)
PCR	Hcp 100	FFPE	7 (7 *Histoplasma*-positive biopsies)	100	100	Moreno-Coutino et al. (2015)
Nested PCR	Hcp 100	Respiratory samples	40 (3 *Histoplasma*-positive samples)	100	100	Almeida-Silva (2015)
PCR	RYP1	Blood	21 (15 from proven histoplasmosis patients[a])	87	100	Brilhante et al. (2016)
Nested PCR	NAALADase	Serum, FFPE, BAL fluid	9 (5 from proven or probable histoplasmosis patients)	77	NR	Muraosa et al. (2016)
Real-time PCR	NAALADase	Serum, FFPE, BAL fluid	9 (5 from proven or probable histoplasmosis patients)	33	NR	Muraosa et al. (2016)
Nested PCR	Hcp 100	Serum	7 (7 *Histoplasma*-positive serology)	86	NR	Frias-De-Leon et al. (2017)
Simplex PCR	Specific gene	Serum	7 (7 *Histoplasma*-positive serology)	86	NR	Frias-De-Leon et al. (2017)

Abbreviations: FFPE, formalin-fixed paraffin-embedded tissue; GAPDH, Glyceraldehyde 3-phosphate dehydrogenase; NAALADase, *N*-acetyl-aspartyl-glutamate peptidase; BAL, bronchoalveolar lavage fluid; NR, not reported.
[a] EORTC criteria.

A molecular assay, non–PCR-based, loop-mediated isothermal amplification (LAMP) was utilized to detect Hcp100 in clinical isolates and DNA obtained from urine of patients with proven histoplasmosis. This method showed 100% sensitivity and specificity in detecting DNA extracted from isolates and 67% sensitivity in detecting *Histoplasma* DNA in urine samples. LAMP is cheaper than traditional PCR, but more study will be necessary for a better evaluation of this methodology (Scheel et al., 2014).

12.7 Summary and conclusions

Histoplasmosis continues to cause significant morbidity and mortality, especially in the setting of increasing numbers of immunocompromised hosts. Despite the scientific advances in the study of *H. capsulatum*, major difficulties persist in our capacity to rapidly diagnose disease due to this fungus. A great effort in the search for the development of new methodologies for the more accurate diagnosis of histoplasmosis has been made, since most of the diagnostic methods present significant limitations, due to false positive and false negative results. Reagents such as purified or recombinant antigens and monoclonal antibodies have been developed to attempt to improve serological tests. New targets and enhancement of molecular methodologies are being studied, as well as new standardized protocols. However, all tests, be they serological or molecular, need to be validated, with adequate accuracy, linearity, detection threshold and quantification, specificity, reproducibility, and stability. However, the gold standard for diagnosis is still the isolation of *H. capsulatum* on culture. Selecting the appropriate tests requires an understanding of the performance characteristics of various diagnostic methods in each clinical setting.

References

Almeida, M. A. 2015. Validação de ensaio imunoenzimático (Western blot) para o diagnóstico da histoplasmose. Msc diss., Fiocruz.

Almeida, M. A., Pizzini, C. V., Damasceno, L. S. et al. 2016. Validation of Western blot for *Histoplasma capsulatum* antibody detection assay. *BMC Infect Dis* 16: 87.

Almeida-Silva, F. 2015. Desempenho de métodos laboratoriais para o diagnóstico de pacientes com suspeita de pneumonia por *pneumocystis jirovecii* no Instituto Nacional de Infectologia Evandro Chagas. Msc diss., Fiocruz.

Azar, M. M., Hage, C. A. 2017. Laboratory diagnostics for histoplasmosis. *J Clin Microbiol* 55: 1612–1620.

Babady, N. E., Buckwalter, S. P., Hall, L. et al. 2011. Detection of *Blastomyces dermatitidis* and *Histoplasma capsulatum* from culture isolates and clinical specimens by use of real-time PCR. *J Clin Microbiol* 49: 3204–3208.

Bialek, R., Feucht, A., Aepinus, C. et al. 2002. Evaluation of two nested PCR assays for detection of *Histoplasma capsulatum* DNA in human tissue. *J Clin Microbiol* 40: 1644–1647.

Bracca, A., Tosello, M. E., Girardini, J. E. 2003. Molecular detection of *Histoplasma capsulatum* var. capsulatum in human clinical samples. *J Clin Microbiol* 41:1753–1755.

Brilhante, R. S. N., Guedes, G. M. M., Riello, G. B. et al. 2016. RYP1 gene as a target for molecular diagnosis of histoplasmosis. *J Microbiol Methods* 130: 112–114.

Buitrago, M. J., Berenguer, J., Mellado, E. et al. 2006. Detection of imported histoplasmosis in serum of HIV-infected patients using a real-time PCR-based assay. *Eur J Clin Microbiol Infect Dis* 25: 665–668.

Caceres, D. H., Scheel, C. M., Tobón, A. M. et al. 2014. Validation of an enzyme-linked immunosorbent assay that detects *Histoplasma capsulatum* antigenuria in Colombian patients with AIDS for diagnosis and follow-up during therapy. *Clin Vaccine Immunol* 21: 1364–1368.

Couppié, P., Aznar, C., Carme, B. et al. 2006. American histoplasmosis in developing countries with a special focus on patients with HIV: Diagnosis, treatment, and prognosis. *Curr Opin Infect Dis* 19: 443–449.

Dantas, K. C., Moreira, A. P. V., Bernard, G. et al. 2013. The use of nested polymerase chain reaction (nested PCR) for the early diagnosis of *Histoplasma capsulatum* infection in serum and whole blood of HIV-positive patients. *An Bras Dermatol* 88: 141–143.

Falci, D. R., Hoffmann, E. R., Paskulin, D. D. 2017. Progressive disseminated histoplasmosis: A systematic review on the performance of non-culture-based diagnostic tests. *Braz J Infect Dis* 21: 7–11.

Fressatti, R., Dias-Siqueira, V. L., Svidzinski, T. I. 1992. A medium for inducing conversion of *Histoplasma capsulatum* var. *capsulatum* into its yeast-like form. *Mem Inst Oswaldo Cruz* 87: 53–58.

Frías-De-Leon, M. G., Ramírez-Bárcenas, J. A., Rodríguez-Arellanes, G. et al. 2017. Usefulness of molecular markers in the diagnosis of occupational and recreational histoplasmosis outbreaks. *Folia Microbiol* 62: 111–116.

Guimarães, A. J., Nosanchuk, J. D., Zancopé-Oliveira, R. M. 2006. Diagnosis of histoplasmosis. *Braz J Microbiol* 37: 1–13.

Hage, C. A., Ribes, J. A., Wengenack, N. L. et al. 2011. A multicenter evaluation of tests for diagnosis of histoplasmosis. *Clin Infect Dis* 53: 448–454.

Irinyl, L., Serena, C., Garcia-Hermoso, D. 2015. International Society of Human and Animal Mycology (ISHAM)-ITS reference DNA barcoding database—The quality controlled standard tool for routine identification of human and animal pathogenic fungi. *Med Mycol* 53: 313–337.

Kozel, T. R., Wickes B. 2014. *Fungal diagnostics. Cold Spring Harb Perspect Med* 4: a019299.

Lynch Jr., H. J., Plexico, K. L. 1962. A rapid method for screening sputums for *Histoplasma capsulatum* employing the fluorescent-antibody technic. *N Engl J Med* 266: 811–814.

Maubon, D., Simon, S., Aznar, C. 2007. Histoplasmosis diagnosis using a polymerase chain reaction method. Application on human samples in French Guiana, South America. *Diag Microbiol Infect Dis* 58: 441–444.

Moraes, A. M. L., Almeida-Paes R., Morandi, V. L. 2009. Micologia. In *Conceitos e métodos para a formação de profissionais em laboratórios de saúde*, ed. E. M. Molinaro, L. F. G. Caputo, and M. R. R. Amendoeira v. 4: 399–496. Rio de Janeiro: EPSJV; IOC.

Moreno-Coutiño, G., Hernández-Castro, R., Toussaint-Cair, S. et al. 2015. Histoplasmosis and skin lesions in HIV: A safe and accurate diagnosis. *Mycoses* 58: 413–415.

Muñoz, C., Gómez, B. L., Tobón, A. et al. 2010. Validation and clinical application of a molecular method for identification of *Histoplasma capsulatum* in human specimens in Colombia, South America. *Clin Vaccine Immunol* 17: 62–67.

Muraosa, Y., Toyotome, T., Yahiro, M. et al. 2016. Detection of *Histoplasma capsulatum* from clinical specimens by cycling probe-based real-time PCR and nested real-time PCR. *Med Mycol* 54: 433–438.

Nacher, M., Adenis, A., Arathoon, E. et al. 2016. Disseminated histoplasmosis in Central and South America, the invisible elephant: The lethal blind spot of international health organizations. *AIDS* 30: 167–170.

Ohno, H., Tanabe, K., Umeyama, T. et al. 2013. Application of nested PCR for diagnosis of histoplasmosis. *J Infect Chemother* 19: 999–1003.

Pizzini, C. V., Zancopé-Oliveira, R. M., Reiss, E. et al. 1999. Evaluation of a Western blot test in an outbreak of acute pulmonary histoplasmosis. *Clin Diagn Lab Immunol* 6: 20–23.

Powers-Fletcher, M. V., Hanson, K. E. 2016. Nonculture diagnostics in fungal disease. *Infect Dis Clin North Am* 30: 37–49.

Scheel, C., Gómez, B. 2014. Diagnostic methods for histoplasmosis: Focus on endemic countries with variable infrastructure levels. *Curr Trop Med Rep* 1: 129–137.

Scheel, C. M., Zhou, Y., Theodoro, R. C. et al. 2014. Development of a loop-mediated isothermal amplification method for detection of *Histoplasma capsulatum* DNA in clinical samples. *J Clin Microbiol* 52: 483–488.

Silva, R. M., Silva Neto, J. R., Santos, C. S. et al. 2015. Fluorescent *in situ* hybridization of pre-incubated blood culture material for rapid diagnosis of histoplasmosis. *Med Mycol* 53: 160–164.

Simon, S., Vern, V., Boukhari, R. et al. 2010. Detection of *Histoplasma capsulatum* DNA in human samples by real-time polymerase chain reaction. *Diag Microbiol Infect Dis* 66: 268–273.

Tang, Y. W., Durkin, M. M., Sefers, S. E. et al. 2006. Urine polymerase chain reaction is not as sensitive as urine antigen for the diagnosis of disseminated histoplasmosis. *Diag Microbiol Infect Dis* 54: 283–287.

Theel, E. S., Harring, J. A., Dababneh, A. S. et al. 2015. Reevaluation of commercial reagents for detection of *Histoplasma capsulatum* antigen in urine. *J Clin Microbiol* 53: 1198–1203.

Toranzo, A. I., Tiraboschi, I. N., Fernández, N. et al. 2007. Molecular diagnosis of human histoplasmosis in whole blood samples. *Rev Argent Microbiol* 41: 20–26.

Valero, C., Buitrago, M. J., Gago, S. et al. 2017. A matrix-assisted laser desorption/ionization time of flight mass spectrometry reference database for the identification of *Histoplasma capsulatum*. *Med Mycol* 56: 307–314.

Wheat, J., French, M. L., Kohler, R. B. et al. 1982. The diagnostic laboratory tests for histoplasmosis: Analysis of experience in a large urban outbreak. *Ann Intern Med* 97: 680–685.

Wheat, L. J., Azar, M. M., Bahr, N. C. 2016. Histoplasmosis. *Infect Dis Clin North Am* 30: 207–227.

Wheat, L. J., Kohler, R., Tewari, R. 1986. Diagnosis of disseminated histoplasmosis by detection of *Histoplasma capsulatum* antigen in serum and urine specimens. *N Engl J Med* 314: 83–88.

Zancopé-Oliveira, R. M., Pizzini, C. V., Muniz, M. M. et al. 2014. Diagnostic aspects of paracoccidioidomycosis. *Curr Trop Med Rep* 1: 111–118.

Zhang, X., Gibson, B., Daly, T. M. 2013. Evaluation of commercially available reagents for diagnosis of histoplasmosis infection in immunocompromised antigen from urine specimens. *J Clin Microbiol* 51: 4095–4101.

13
Paracoccidioides Complex

Zoilo Pires de Camargo and Anderson Messias Rodrigues

Contents

13.1 Introduction

Paracoccidioidomycosis (PCM) is a mycotic disease caused by species of the genus *Paracoccidioides*, a group of thermally dimorphic fungi that grow in a mycelial form at room temperature and as budding yeasts when cultured at 37°C or in parasitism in the host tissues. PCM is limited to Latin American countries, and the most important regions of endemicity are found in Brazil, Colombia, and Venezuela (Martinez, 2017). Paracoccidioidomycosis is considered a neglected tropical disease (Queiroz-Telles et al., 2017) that affects occupational groups routinely dealing with soil, such as farmers, saw millers, and rural workers, especially coffee pickers in South America. Paracoccidioidomycosis develops after environmental exposure following inhalation of infective propagules, which proliferate and settle in the lungs of the host. From the lungs, the fungus can disseminate to other organs and systems by hematogenic or lymphatic pathway (Bocca et al., 2013). PCM presents as two major clinical forms: the acute or sub-acute form and the chronic form (da Costa & Silva, 2014).

In Brazil, PCM is considered the eighth most common cause of death among infectious and parasitic chronic diseases, with a mortality rate of 1.45 per million population (Coutinho et al., 2002). Epidemiological data from 3181 deaths from PCM in Brazil during 16 years of sequential data (from 1980 to 1995) revealed a considerable magnitude and low visibility of PCM, representing the eighth most common cause of death from predominantly chronic or recurrent types of infectious and parasitic diseases (Coutinho et al., 2002). Such a mortality rate justifies classifying this disease as an important health problem in Brazil. However, no government programs exist for this mycosis, with rare punctual exceptions.

For over a century, PCM was thought to be caused solely by *P. brasiliensis*, but recent studies (Carrero et al., 2008; Desjardins et al., 2011; Matute et al., 2006a, b; Teixeira et al., 2009; Theodoro et al., 2012) have shown that other species of *Paracoccidioides* can also cause PCM. Molecular and morphological studies of clinical and environmental isolates of etiological agents of PCM led to the recognition of two distinct biological species: *P. brasiliensis sensu lato* (containing at least four cryptic species: S1, PS2, PS3, and PS4) and *P. lutzii* (originally named Pb01-like) (Carrero et al., 2008; Desjardins et al., 2011; Matute et al., 2006a, b; Teixeira et al., 2009; Theodoro et al., 2012). As observed in phylogenetic studies, molecular siblings in *Paracoccidioides* show different biology and geographic distribution. So far, S1 isolates are widely distributed throughout Latin America, occurring in Argentina, Brazil, Paraguay, Peru, and

Venezuela; PS2 is found in Argentina, Brazil, Venezuela, and Uruguay; PS3 occurs in Colombia, Venezuela, and Brazil; PS4 isolates have thus far been described from Venezuela; and *P. lutzii* is found in the central-west and north regions of Brazil and Ecuador (Hahn et al., 2014; Marques-da-Silva et al., 2012; Roberto et al., 2016; Theodoro et al., 2012).

Recently, it was proposed to elevate the cryptic species in the *P. brasiliensis* complex to species status, that is, *P. brasiliensis sensu stricto* (formerly S1), *P. americana sp. nov.* (formerly PS2), *P. restrepiensis sp. nov.* (formerly PS3), and *P. venezuelensis sp. nov.* (formerly PS4) (Turissini et al., 2017), in addition to the previously described *P. lutzii* (Teixeira et al., 2009, 2014). Whether these taxonomic novelties represent relevant aspects in clinical settings remains to be investigated in depth, associating molecular epidemiology and clinical data in light of a consilient species concept (Hahn et al., 2019).

Notwithstanding, the differentiation between *P. brasiliensis s.l.* and *P. lutzii* is relevant under the serology of PCM. Several studies have demonstrated that false positive results using double immune diffusion tests were mainly due to an incorrect use of antigens from the genus, as only antigens of *P. brasiliensis* (PS3) have been used for this purpose. Shortly after the discovery of *P. lutzii*, it became clear that new antigens from *P. lutzii* (including antigens derived from autochthone isolates) would be needed to obtain more accurate diagnoses (Batista Jr. et al., 2010; Gegembauer et al., 2014; Queiroz Junior et al., 2014).

13.2 Mycological diagnosis of paracoccidioidomycosis

A clinical diagnosis of PCM needs to be confirmed through laboratory exams. Direct microscopic methods, such as wet mounts using KOH and conventional histopathology, provide valuable information that quickly leads to diagnosis of PCM due to the presence of pathognomonic forms of *Paracoccidioides* spp. (usually referred as "steering wheel" or "Mickey Mouse" shapes), which are unique among the group of endemic fungi that cause systemic mycoses. In KOH preparations (Figure 13.1a and b) or Gomori-stained tissues, *Paracoccidioides* cells tend to be seen undergoing multiplication and, often, producing multiple buds (Restrepo, 2000), a gold standard for PCM diagnostic (Figure 13.1i). HE stain can be used; however, the structures are more difficult to visualize (Figure 13.1g and h). However, confirmation of microscopic findings by culture is desirable for definitive identification of *Paracoccidioides* (Figure 13.1d through f). It is important to note that direct microscopic methods do not allow differentiating among cryptic *P. brasiliensis* species (S1, PS2, PS3, or PS4) or even between *P. brasiliensis s.l.* and *P. lutzii*. Indirect serological methods such as the detection of antigens and specific antibodies circulating in biological fluids (serum, bronchoalveolar lavage, cerebrospinal fluid) can also lead to a presumptive diagnosis, but no species can be recognized. In this scenario, molecular assays based on the nucleic acid sequencing of *Paracoccidioides*-specific targets or selective amplification of DNA targets using the polymerase chain reaction provides better, faster laboratory diagnosis of PCM; however, traditional methods such as microscopy and culture remain primary and essential tools.

13.3 Collection, transport, and processing of clinical specimens

The first step in acquiring a quality mycological diagnosis result for any patient is the specimen collection procedure. The collection and transport of clinical material depend on several factors from the site and type of injury, type of affected tissue, and patient's condition. One should collect a large quantity of material for examination. The material must be properly identified, accompanied by the medical application with patient data indications, clinical suspicion, underlying diseases if any, and the test requested. The transport of material must be made in suitable sterile containers, and the collection of clinical material and processing should be performed in a timely manner to prevent deterioration of clinical material to be used in tests. The above recommendations are decisive and will ensure proper stability of the specimen and more accurate test results.

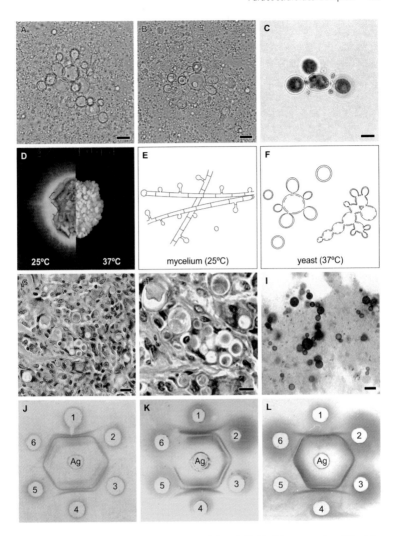

Figure 13.1   *Paracoccidioides* spp. diagnosis: (A) and (B) birefringent and multibudding cells of *Paracoccidioides* spp. in biological material; (C) birefringent and multibudding cells of *Paracoccidioides* spp. from a 37°C culture (yeast form); (D) *Paracoccidioides* spp. culture at room temperature − 25°C, mycelial form (at left), and at 37°C, yeast form (at right); (E) *Paracoccidioides* spp. mycelium form (25°C) (scheme); (F) *Paracoccidioides* spp. yeast form (37°C) (scheme); (G) *Paracoccidioides* spp. in tissue stained by HE (40×); (H) *Paracoccidioides* spp. in tissue stained by HE (100×). (I) *Paracoccidioides* spp. in tissue stained by Grocott method (bars = 10 µm); (J, K, L) examples of immunodiffusion tests for paracoccidioidomycosis, in the central well, *Paracoccidioides brasiliensis* antigen (Ag) and in the peripherical wells different patient sera.

13.4 Direct examination

Sputum has a long and rich history as a source for searching for *Paracoccidioides* spp. yeast cells and remains an important sample for laboratorial diagnosis. Early-morning specimens of sputum are best, with prior cleaning of the mouth with water using two or three mouth washes. It is recommended to collect at least three sputum samples on alternate days, if possible. Sputum should be stored in a sterile wide-mouth jar and transported in boxes suitable for transportation, often with small containers with ice or icy water. Also, bronchial and bronchoalveolar washings are transported in sterile screw-cap tubes or jars. On reaching the laboratory, 4% NaOH or N-acetyl-L-cysteine and dithiothreitol must be added to the sputum to liquefy it, and the incubating temperature should be left at 37°C for 2–3 hours. When the sputum is liquefied, it should be centrifuged and the supernatant discarded. With a Pasteur pipette type, a drop of the precipitate is placed on a microscope slide and a cover slip is placed over it. Then, the precipitate is examined under a light microscope for birefringent and multibudding yeasts (Figure 13.1a through c).

In the case of mucosal scrapings, skin lesions, or pus, the material must be placed on a glass slide with a drop of 10% KOH, with or without the use of Calcofluor white and a coverslip. It must be examined under a microscope with 40× magnification and searched for birefringent multibudding yeast cells. However, if only single cells or chains of cells are present during direct examination, the microorganism cannot be differentiated from other fungal pathogens. Direct examination of sputum and other secretions does not allow differentiating among the *Paracoccidioides* species described to date. In this scenario, it is necessary to use molecular data (PCR-RFLP, DNA sequencing, species-specific PCR) to recognize cryptic entities. As with most laboratory procedures, positive findings are of greater value than negative ones; detection of fungal elements can confirm the diagnosis of PCM, but negative microscopy does not exclude it.

13.5 Histopathological examination

Biopsy material should be divided for separate histological and cultural studies. Biopsy is the removal of tissue from any part of the body to examine it for disease. Some may remove a small tissue sample with a needle, while others may surgically remove a suspicious nodule or lump. Most needle biopsies are performed on an outpatient basis with minimal preparation. The physician must provide instructions based on the type of biopsy being performed.

In the case of suspected PCM, the biopsy may be obtained from localized lesions on the skin, oral or anal mucosa, pulmonary puncture, or any other site. The obtained fragment should be placed in a 10% formalin flask and sent to the laboratory for paraffin embedding and making fine cuts for fungal-specific staining. A huge range of stains are available for staining, such as periodic acid-Schiff reagent (PAS) and silver impregnation (Grocott/Gomori) (Teixeira et al., 1978). Histologically, some giant cells and many *Paracoccidioides* yeast-forms can be observed; these are free or inside the macrophages.

13.6 Cultural examination

The definitive diagnosis of PCM depends on isolation of *Paracoccidioides in vitro* (Richardson & Warnock, 2012). All PCM-suspect clinical material should be cultured in several culture media for fungi and under ideal conditions for growth. In mycology, isolation and identification of fungi from clinical material can usually be achieved with traditional cultural procedures. The basic culture medium is Sabouraud Dextrose agar; many are supplemented with antibiotics such as penicillin and streptomycin, chloramphenicol, or gentamicin to prevent the growth of contaminating bacteria in samples. However, if it is necessary to prevent the growth of contaminating fungi, such as anemophilous fungi, cycloheximide (Actidione) can be added to the medium. After cultivating the clinical material in several tubes, they should be incubated at both room temperature and at 35°C–37°C. *Paracoccidioides* is a fastidious fungus, taking up to 30 days to grow at room temperature (Figure 13.1d, left). However, at elevated temperatures (35°C–37°C), *Paracoccidioides* might grow faster, and after 7–10 days, yeast colonies, usually cream and cerebriform, can be observed (Figure 13.1d, right). Isolates of *Paracoccidioides* spp.

at 25°C–30°C grow slowly and produce colonies that vary in gross morphology, ranging from glabrous, brown colonies to wrinkled, floccose, beige or white colonies (Restrepo & Correa, 1972). Conidia at 24°C are absent or are intercalary-like chlamydospore cells (de Hoog et al., 2000), but not a distinctive feature of the cryptic species in the *P. brasiliensis* complex (Theodoro et al., 2012). The mold form of *Paracoccidioides* spp. may require growth for several weeks, and conversion to the yeast form is necessary for definitive identification, when multiply budding yeast cells are visualized.

13.7 Serological diagnosis

Tests for the detection of antibodies against *P. brasiliensis* are useful and were first used as a diagnostic aid in the 1960s (Fava-Netto, 1961), although after decades of studies, the establishment of an ideal assay for PCM presenting high sensitivity and specificity was not found. The existence of cross-reacting antigens among the dimorphic fungi, due to the complexity of the fungal cell wall, limits the specificity of the serological reactions. Furthermore, antigenic preparations obtained from *P. brasiliensis* currently used were developed empirically. Exceptionally, the immunodiffusion test for PCM has proven very useful in routine laboratory so far. Most antigenic preparations used for testing are derived from crude culture filtrates containing a range of metabolic products (de Camargo et al., 1988). In addition, purified antigens such as glycoprotein of 43,000 daltons (gp43) and other purified molecules have also been used for serological diagnosis. Other assays such as ELISA (Albuquerque et al., 2005), dot blot (Taborda & Camargo, 1994), and Western blot (Camargo et al., 1989) may be used as adjuncts for diagnosis. More recently, a simple slide agglutination test using latex particles was proposed for diagnostic purposes, for detection of antigens and antibodies circulating in biological fluids of patients, with excellent results (Dos Santos et al., 2015). The serological tests above are not limited to diagnosis but are also useful to monitor the patient response to therapy (Marques-da-Silva et al., 2004; da Silva et al., 2004).

Regarding the serology of PCM due to *P. lutzii*, an ideal antigen for use in immunodiffusion tests is the antigenic preparation known as "cell-free antigen," or CFA, obtained from molecules present on the surface of yeasts (Gegembauer et al., 2014).

To date, there are no serological tests to differentiate PCM caused by different species of the *P. brasiliensis* complex. A strategy has been put forth in our laboratories: all sera from patients suspected of PCM due to *P. brasiliensis* or *P. lutzii* should be tested against the traditional exoantigen of *P. brasiliensis* (AgPbB339) and purified *gp43* molecules derived from the same standard strain. The reactive sera are then considered to be PCM due to *P. brasiliensis* (Figure 13.1j through l). The non-reactive sera are then tested against cell-free antigen from *P. lutzii*; positive reactivity with this CFA antigen leads to the conclusion of a PCM due to *P. lutzii*.

13.8 Molecular diagnosis

Phylogenetically related *Paracoccidioides* species are often misidentified due to overlapping phenotypic characteristics, and therefore morphological features alone are inadequate to differentiate among S1, PS2, PS3, PS4, and *P. lutzii*. Molecular techniques are considered the best means of distinguishing among cryptic groups in *Paracoccidioides*. DNA sequencing followed by phylogenetic analysis is a gold-standard method for species recognition (Roberto et al., 2016) (Figure 13.2). A great variety of polymerase chain reaction (PCR)-based molecular techniques—such as random amplified polymorphic DNA-PCR (Hahn et al., 2003), species-specific PCR (Alves et al., 2015), nested PCR (Arantes et al., 2013), microsatellite markers (Matute et al., 2006b), the SNaPshot method (Theodoro et al., 2012), and real-time PCR (Buitrago et al., 2009)—have been investigated as possible alternatives for routine identification of *Paracoccidioides*. In addition, matrix-assisted laser desorption ionization time-of-flight (MALDI-ToF) is an important aid in differentiating *P. brasiliensis* complex and *P. lutzii*, but no species embedded in the *P. brasiliensis* complex (S1, PS2, PS3, and PS4) can be discriminated (de Almeida et al., 2015).

Diagnosis of PCM based on molecular procedures is very promising, since it allows fungal detection from contaminated samples as from isolated strains. The nucleotide sequence derived from the major *Paracoccidioides* antigen stands out as the primary marker for molecular

Multilocus sequence analysis (*ARF+GP43+TUB1*)

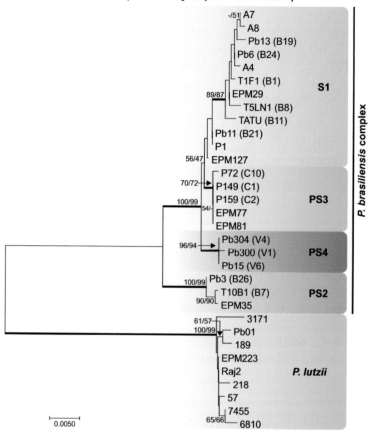

Figure 13.2   Phylogenetic relationships between members of the genus *Paracoccidioides* based on GP43, ADP-ribosylation factor (ARF), and TUB1 partial genes. Neighbor-joining (NJ) and maximum likelihood (ML) analyses were performed using a Kimura two-parameter model. The percentage of replicate trees in which the associated taxa clustered together in the bootstrap test (1000 replicates) is shown next to the branches (NJ/ML). The clades correspond to the genetic groups *P. lutzii* and S1, PS2, PS3, and PS4 of *P. brasiliensis*. Bar = total nucleotide differences between taxa.

identification and genetic diversity studies (Morais et al., 2000). Cisalpino et al. (1996) cloned and characterized the entire coding region of the *gp43* gene, and subsequently, Gomes et al. (2000) determined various primer pairs and suggested a PCR for direct amplification of *Paracoccidioides* DNA from sputum or nested PCR. Bialek et al. (2000) proposed a nested PCR based on the *gp43* gene for identification of *P. brasiliensis* in tissue sections. Ricci et al. (2008) identified *P. brasiliensis* in tissue sections from PCM patients, but with low positivity (30%), using the same primers described by Bialek et al. (2000).

DNA sequencing can also be used for diagnosis of PCM, and the internal transcribed spacers 1 and 2 (ITS1/2) in the rDNA work as a barcoding marker (Irinyi et al., 2015; Hebeler-Barbosa et al., 2003). Motoyama et al. (2000) amplified and sequenced the 5.8S and 28S ribosomal DNA genes and intergenic regions of *P. brasiliensis* and, using primers specifically designed for both ribosomal DNA regions, were able to discriminate between *P. brasiliensis* and other human pathogenic fungi by PCR. The use of this molecular marker could be important for PCM diagnosis and ecological and molecular epidemiological studies.

P. brasiliensis contains several different cryptic species, as observed by phylogenetic studies using multilocus sequence analysis (MLSA) (Carrero et al., 2008; Matute et al., 2006a) (Figure 13.2), microsatellite analysis (Matute et al., 2006b), and *PRP8* intein sequencing (Theodoro et al., 2008). Matute et al. (2006b) analyzed eight regions of five nuclear coding genes: chitin synthase, β-glucan synthase, α-tubulin, adenyl ribosylation factor, and PbGP43, and performed microsatellite analysis that revealed three distinct, previously unrecognized species: S1 (species 1 from Brazil, Argentina, Paraguay, Peru, and Venezuela), PS2 (phylogenetic species 2 from Brazil and Venezuela), and PS3 (phylogenetic species from Colombia). Recently, the recognition of all five species in the *Paracoccidioides* complex to a taxonomic species status was proposed; that is, *P. brasiliensis sensu stricto* (S1), *P. americana sp. nov.* (PS2), *P. restrepiensis sp. nov.* (PS3), *P. venezuelensis sp. nov.* (PS4), and *P. lutzii* (Turissini et al., 2017) (Figure 13.2). However, a consilient species concept is mandatory in modern taxonomy, based on the convergence of multiple, independent data sets, as a means of delimiting species (Jančič et al., 2015). In the clinical setting, the need for recognition of *P. americana*, *P. restrepiensis*, and *P. venezuelensis* as entities different from *P. brasiliensis* remains uncertain until classified into a consilient context. The most sensible scenario supports the differentiation between *P. brasiliensis* and *P. lutzii* based on evidence originating from serological (Gegembauer et al., 2014; Queiroz Junior et al., 2014; Machado et al., 2013) and epidemiological divergences (Theodoro et al., 2012).

Loop mediated isothermal amplification (LAMP) is a single-tube technique for the amplification of DNA and was successfully used to detect in a specific manner the *gp43* gene of *P. brasiliensis* (Endo et al., 2004) in a set of 22 clinical and 7 armadillo-derived isolates.

Buitrago et al. (2009) developed real-time PCR for the detection of *P. brasiliensis* DNA in the diagnosis of paracoccidioidomycosis in patient clinical samples and identification in cultures. Rocha-Silva et al. (2017) developed and standardized rt-PCR, opening perspectives to molecular diagnosis development for paracoccidioidomycosis, since rt-PCR can be applied to a broad spectrum of infectious diseases. It would need to be tested in biological samples in order to validate this method and then generate a diagnostic kit for paracoccidioidomycosis. Rocha-Silva et al. (2018) used rt-PCR for diagnosis of PCM, previously diagnosed as neoplasm.

Roberto et al. (2016) proposed identifying cryptic *Paracoccidioides* spp. using polymerase chain reaction restriction fragment length polymorphism (PCR-RFLP) of the alpha-tubulin (*TUB1*) gene. *In silico* analysis of 90 *TUB1* sequences led to the identification of two restriction enzymes with the potential to identify *Paracoccidioides*: *Bcl* I and *Msp* I. A portion of the *TUB1* gene was amplified and double digested *in vitro* with the *Bcl* I and *Msp* I endonucleases, which generated four different electrophoretic patterns corresponding to the four main genetic groups: S1, PS2, and PS3 of *P. brasiliensis* and *P. lutzii*. These data showed that *TUB1*-RFLP could efficiently discriminate among distinct *Paracoccidioides* species with high accuracy.

In summary, it is important to note that not all molecular tests available allow differentiating among all *Paracoccidioides* spp., and in many cases, the combination of two or more techniques is recommended to achieve highly discriminatory results.

Acknowledgments

The authors acknowledge the financial support of São Paulo Research Foundation (FAPESP 2017/27265-5 and 2018/21460-3), National Council for Scientific and Technological Development (CNPq 429594/2018-6) and Coordination for the Improvement of Higher Education Personnel (CAPES 88887.177846/2018-00). The funders had no role in the study design, data collection and analysis, decision to publish or preparation of the manuscript.

References

Albuquerque, C. F., Marques da Silva, S. H., Camargo, Z. P. 2005. Improvement of the specificity of an enzyme-linked immunosorbent assay for diagnosis of paracoccidioidomycosis. *J Clin Microbiol* 43: 1944–1946.

Alves, F. L., Mariceli, A. R., Hahn, R. C. et al. 2015. Transposable elements and two other molecular markers as typing tools for the genus *Paracoccidioides*. *Med Mycol* 53: 165–170.

Arantes, T. D., Theodoro, R. C., Macoris, S. A. G., Bagagli, E. 2013. Detection of *Paracoccidioides* spp. in environmental aerosol samples. *Med Mycol* 51: 83–92.

Batista Jr., J., Camargo, Z. P., Fernandes, G. F., Vicentini, A. P., Fontes, C. J. F. F., Hahn, R. C. 2010. Is the geographical origin of a Paracoccidioides brasiliensis isolate important for antigen production for regional diagnosis of paracoccidioidomycosis? *Mycoses* 53: 176–180.

Bialek, R., Ibricevic, A., Aepinus, C. et al. 2000. Detection of *Paracoccidioides brasiliensis* in tissue samples by a nested PCR assay. *J Clin Microbiol* 38: 2940–2942.

Bocca, A. L., Amaral, A. C., Teixeira, M. M., Sato, P. K., Shikanai-Yasuda, M. A., Soares Felipe, M. S. 2013. Paracoccidioidomycosis: Eco-epidemiology, taxonomy and clinical and therapeutic issues. *Future Microbiol* 8: 1177–1191.

Buitrago, M. J., Merino, P., Puente, S. et al. 2009. Utility of real-time PCR for the detection of *Paracoccidioides brasiliensis* DNA in the diagnosis of imported paracoccidioidomycosis. *Med Mycol* 47: 879–882.

Camargo, Z. P., Unterkircher, C., Travassos, L. R. 1989. Identification of antigenic polypeptides of *Paracoccidioides brasiliensis* by immunoblotting. *J Med Vet Mycol* 27: 407–412.

Carrero, L. L., Nino-Vega, G., Teixeira, M. M. et al. 2008. New *Paracoccidioides brasiliensis* isolate reveals unexpected genomic variability in this human pathogen. *Fungal Genet Biol* 45: 605–612.

Cisalpino, P., Puccia, R., Yamauchi, L. M., Cano, M. I., da Silveira, J. F., Travassos, L. R. 1996. Cloning, characterization, and epitope expression of the major diagnostic antigen of *Paracoccidioides brasiliensis*. *J Biol Chem* 271: 4553–4560.

Coutinho, Z. F., Da Silva, D., Lazera, M. et al. 2002. Paracoccidioidomycosis mortality in Brazil (1980–1995). *Cad Saude Publica* 18: 1441–1454.

da Costa, M. M., Silva, H. M. S. 2014. Epidemiology, clinical, and therapeutic aspects of paracoccidioidomycosis. *Curr Trop Med Rep* 1: 138–44.

da Silva, S. H. M., Queiroz-Telles, F., Colombo, A. L., Blotta, M. H. S. L., Lopes, J. D., Camargo, Z. P. 2004. Monitoring gp43 antigenemia in paracoccidioidomycosis patients during therapy. *J Clin Microbiol* 42: 2419–2424.

de Almeida, J. N., Del Negro, G. M. B., Grenfell, R. C. et al. 2015. MALDI-TOF mass spectrometry for rapid identification of the dimorphic fungi *Paracoccidioides brasiliensis* and *Paracoccidioides lutzii*. *J Clin Microbiol* 53: 1383–1386.

de Camargo, Z. P., Unterkircher, C., Campoy, S. P., Travassos, L. R. 1988. Production of *Paracoccidioides brasiliensis* exoantigens for immunodiffusion tests. *J Clin Microbiol* 26: 2147–2751.

de Hoog, G. S., Guarro, J., Gené, J., Figueras, M. J. 2000. *Atlas of Clinical Fungi*. 2nd ed. Utrecht, The Netherlands: Centraalbureau voor Schimmelcultures.

Desjardins, C. A., Champion, M. D., Holder, J. W. et al. 2011. Comparative genomic analysis of human fungal pathogens causing paracoccidioidomycosis. *PLoS Genet* 7: e1002345.

Dos Santos, P. O., Rodrigues, A. M., Fernandes, G. F., da Silva, S. H., Burger, E., Camargo, Z. P. 2015. Immunodiagnosis of paracoccidioidomycosis due to *Paracoccidioides brasiliensis* using a latex test: Detection of specific antibody anti-gp43 and specific antigen gp43. *PLoS Negl Trop Dis* 9: e0003516.

Endo, S., Komori, T., Ricci, G. et al. 2004. Detection of gp43 of *Paracoccidioides brasiliensis* by the loop-mediated isothermal amplification (LAMP) method. *FEMS Microbiol Lett* 234: 93–97.

Fava-Netto, C. 1961. Contribuição para o estudo imunológico da blastomicose de Lutz (blastomicose sul-americana). *Rev Inst A Lutz* 21: 99–194.

Gegembauer, G., Araujo, L. M., Pereira, E. F. et al. 2014. Serology of paracoccidioidomycosis due to *Paracoccidioides lutzii*. *PLoS Negl Trop Dis* 8: e2986.

Gomes, G. M., Cisalpino, P. S., Taborda, C. P., Camargo, Z. P. 2000. PCR for diagnosis of paracoccidioidomycosis. *J Clin Microbiol* 38: 3478–3480.

Hahn, R. C., Macedo, A. M., Fontes, C. J., Batista, R. D., Santos, N. L., Hamdan, J. S. 2003. Randomly amplified polymorphic DNA as a valuable tool for epidemiological studies of *Paracoccidioides brasiliensis*. *J Clin Microbiol* 41: 2849–2854.

Hahn, R. C., Rodrigues, A. M., Fontes, C. J. et al. 2014. Fatal fungemia due to *Paracoccidioides lutzii*. *Am J Trop Med Hyg* 91: 394–398.

Hahn, R.C., Rodrigues, A.M., Della Terra, P.P. et al. 2019. Clinical and epidemiological features of paracoccidioidomycosis due to *Paracoccidioides* lutzii. *PLoS Negl Trop Dis* 13: e0007437.

Hebeler-Barbosa, F., Morais, F. V., Montenegro, M. R. et al. 2003. Comparison of the sequences of the internal transcribed spacer regions and PbGP43 genes of *Paracoccidioides brasiliensis* from patients and armadillos (*Dasypus novemcinctus*). *J Clin Microbiol* 41: 5735–5737.

Irinyi, L., Serena, C., Garcia-Hermoso, D. et al. 2015. International Society of Human and Animal Mycology (ISHAM)-ITS reference DNA barcoding database-the quality controlled standard tool for routine identification of human and animal pathogenic fungi. *Med Mycol* 53: 313–37.

Jančič, S., Nguyen, H. D. T., Frisvad, J. C. et al. 2015. A taxonomic revision of the *Wallemia sebi* species complex. *PLOS ONE* 10: e0125933.

Machado, G. C., Moris, D. V., Arantes, T. D. et al. 2013. Cryptic species of *Paracoccidioides brasiliensis*: Impact on paracoccidioidomycosis immunodiagnosis. *Mem Inst Oswaldo Cruz* 108: 637–643.

Marques da Silva, S. H., Grosso, D. M., Lopes, J. D. et al. 2004. Detection of *Paracoccidioides brasiliensis* gp70 circulating antigen and follow-up of patients undergoing antimycotic therapy. *J Clin Microbiol* 42: 4480–4486.

Marques-da-Silva, S. H., Rodrigues, A. M., de Hoog, G. S., Silveira-Gomes, F., Camargo, Z. P. 2012. Occurrence of *Paracoccidioides lutzii* in the Amazon region: Description of two cases. *Am J Trop Med Hyg* 87: 710–714.

Martinez, R. 2017. New trends in paracoccidioidomycosis epidemiology. *J Fungi* 3: 1.

Matute, D. R., McEwen, J. G., Puccia, R. et al. 2006a. Cryptic speciation and recombination in the fungus *Paracoccidioides brasiliensis* as revealed by gene genealogies. *Mol Biol Evol* 23: 65–73.

Matute, D. R., Sepulveda, V. E., Quesada, L. M. et al. 2006b. Microsatellite analysis of three phylogenetic species of *Paracoccidioides brasiliensis*. *J Clin Microbiol* 44: 2153–2157.

Morais, F. V., Barros, T. F., Fukada, M. K., Cisalpino, P. S., Puccia, R. 2000. Polymorphism in the gene coding for the immunodominant antigen gp43 from the pathogenic fungus *Paracoccidioides brasiliensis*. *J Clin Microbiol* 38: 3960–3966.

Motoyama, A. B., Venancio, E. J., Brandão, G. O. et al. 2000. Molecular identification of *Paracoccidioides brasiliensis* by PCR amplification of ribosomal DNA. *J Clin Microbiol* 38: 3106–3109.

Queiroz Junior, L. P., Camargo, Z. P., Tadano, T. et al. 2014. Serological and antigenic profiles of clinical isolates of *Paracoccidioides* spp. from Central Western Brazil. *Mycoses* 57: 466–472.

Queiroz-Telles, F., Fahal, A. H., Falci, D. R., Caceres, D. H., Chiller, T., Pasqualotto, A. C. 2017. Neglected endemic mycoses. *Lancet Infect Dis* 17: e367–e77.

Restrepo, A. 2000. Morphological aspects of *Paracoccidioides brasiliensis* in lymph nodes: Implications for the prolonged latency of paracoccidioidomycosis? *Med Mycol* 38: 317–322.

Restrepo, A., Correa, I. 1972. Comparison of two culture media for primary isolation of *Paracoccidioides brasiliensis* from sputum. *Sabouraudia* 10: 260–265.

Ricci, G., Zelck, U., Mota, F., Lass-Florl, C., Franco, M. F., Bialek, R. 2008. Genotyping of *Paracoccidioides brasiliensis* directly from paraffin embedded tissue. *Med Mycol* 46: 31–34.

Richardson, M. D., Warnock, D. W. 2012. *Fungal Infection: Diagnosis and Management*. 4th ed. Chichester: Wiley-Blackwell.

Roberto, T. N., Rodrigues, A. M., Hahn, R. C., Camargo, Z. P. 2016. Identifying *Paracoccidioides* phylogenetic species by PCR-RFLP of the alpha-tubulin gene. *Med Mycol* 54: 240–247.

Rocha-Silva, F., Gomes, L. I., Gracielle-Melo, C., Goes, A. M., Caligiorne, R. B. 2017. Real time polymerase chain reaction (rt-PCR): A new patent to diagnostic purposes for paracoccidioidomycosis. *Recent Pat Endocr Metab Immune Drug Discov* 10: 143–149.

Rocha-Silva, F., Guimarães, C. F., Oliveira Júnior, E. R., Figueiredo, S. M., Caligiorne, R. B. 2018. Disseminated paracoccidioidomycosis prediagnosticated as neoplasm: An important challenge in diagnosis using rt-PCR. *Med Mycol Case Rep* 19: 1–5.

Taborda, C. P., Camargo, Z. P. 1994. Diagnosis of paracoccidioidomycosis by dot immunobinding assay for antibody detection using the purified and specific antigen gp43. *J Clin Microbiol* 32: 554–556.

Teixeira, F., Gayotto, L. C., De Brito, T. 1978. Morphological patterns of the liver in South American blastomycosis. *Histopathology* 2: 231–37.

Teixeira, M. M., Theodoro, R. C., Carvalho, M. J. A. et al. 2009. Phylogenetic analysis reveals a high level of speciation in the *Paracoccidioides* genus. *Mol Phylogenet Evol* 52: 273–283.

Teixeira, M. M., Theodoro, R. C., Oliveira, F. F. et al. 2014. *Paracoccidioides lutzii* sp. nov.: Biological and clinical implications. *Med Mycol* 52: 19–28.

Theodoro, R. C., Bagagli, E., Oliveira, C. 2008. Phylogenetic analysis of PRP8 intein in *Paracoccidioides brasiliensis* species complex. *Fungal Genet Biol* 45: 1284–1291.

Theodoro, R. C., Teixeira, M. M., Felipe, M. S. S. et al. 2012. Genus *Paracoccidioides*: Species recognition and biogeographic aspects. *PLoS One* 7: e37694.

Turissini, D. A., Gomez, O. M., Teixeira, M. M., McEwen, J. G., Matute, D. R. 2017. Species boundaries in the human pathogen *Paracoccidioides*. *Fungal Genet Biol* 106: 9–25.

14
Coccidioides

Rossana de Aguiar Cordeiro

Contents

14.1 Introduction: General aspects

Coccidioidomycosis is a deep-seated infection caused by two dimorphic fungal species: *Coccidioides immitis* and *C. posadasii*. Apparently, both species have similar phenotypical and virulence traits, and may cause severe infections with common signs and symptoms. The disease occurs only in the Americas, primarily in areas situated between the north and south 40° latitudes, and the great majority of cases have come from the United States (Hector & Laniado-Laborin, 2005). However, each species shows a specific pattern of distribution: *C. immitis* occurs in central and southern California, and *C. posadasii* occurs in the US states of Nevada, Arizona, New Mexico, and western Texas, as well as in Mexico and Central and South America (Kirkland & Fierer, 2018). *Coccidioides* spp. are found in soils of desert, arid, or semi-arid areas, where they grow saprophytically in the filamentous form. In nearly all cases, the infection results from the inhalation of conidia produced during the filamentous phase, after deliberately stirring the soil—agriculture, hunting, well digging, etc.—or due to environmental disturbances such as sandstorms and earthquakes (Schneider et al., 1997).

During the filamentous phase of the biological cycle, *Coccidioides* spp. form hyaline hyphae that originate arthroconidia up to 7.5 μm in length, intercalated with cells devoid of cytoplasmic material (de Hoog et al., 2000). Arthroconidia are easily detachable from the vegetative mycelium and have cell wall remnants of the breaker cells at their ends, which facilitate their aerial dispersion. Upon being inhaled by a susceptible host, the arthroconidia undergo morphological changes, giving rise to yeast-like structures. In this parasitic phase, large, rounded, thick-walled structures called spherules appear, with a diameter varying from 20–200 μm. These structures increase in size and undergo successive cell divisions, and, upon reaching maturity approximately 72 hours later, release more than 800 endospores—small unicellular structures of 2–5 μm. Each endospore begins the development of a new spherule, resulting in exponential reproduction. Upon reaching the soil, the endospores grow into the filamentous form, thus guaranteeing continuity of the biological cycle (Cox & Magee, 2004).

Coccidioidomycosis can present the following basic clinical forms: asymptomatic, acute pulmonary, chronic pulmonary, disseminated, and primary cutaneous after traumatic inoculation (Ampel, 2010).

Up to 65% of the individuals exposed to infectious arthroconidia do not develop symptoms of the disease, so that infection is only detected in seroepidemiological surveys and positive conversions in intradermal tests with coccidioidin and/or spherulin. Symptomatic infection can occur in approximately 40% of exposed individuals, after an average of 7–21 days of incubation (Ampel, 2010; Cox & Magee, 2004). However, during epidemiological outbreaks following earthquakes, construction work, or excavation at archaeological sites, the rate of symptomatic individuals can reach up to 90% (Schneider et al., 1997).

The acute pulmonary form causes highly diverse clinical symptoms. Patients may develop pneumonia with different radiological patterns, pleural effusion, hilar lymphadenopathy, and pulmonary nodules. The most common clinical syndrome is pneumonia, characterized by cough with or without expectoration, fever, chest pain, headache, muscle fatigue, and anorexia. Approximately 5% of these patients develop erythema nodosum and/or erythema multiforme manifestations resulting from late-type hypersensitivity reactions, which are more frequent in females. The primary symptomatic pulmonary form may regress spontaneously over the course of a few months, even without specific therapy (Cox & Magee, 2004; Malo et al., 2014).

Approximately 5% of patients with primary pneumonia do not achieve spontaneous cure and may develop chronic lung infection, which is manifested by nodular lesions or fibro-cavitary lung disease. Clinically, the disease is characterized by the presence of night sweats, muscle fatigue, weight loss, chronic cough, and hemoptysis (Chiller et al., 2003). Due to the similarity of clinical, radiographic, and histopathological features, progressive pulmonary coccidioidomycosis constitutes an important differential diagnosis with pulmonary tuberculosis (Castañeda-Godoy & Laniado-Laborin, 2002). In Brazil, the disease has been confused with tuberculosis, and reports in the literature attest that despite a negative bacteriological examination, patients were mistakenly submitted to treatment for tuberculosis (Gomes et al., 1978; Moraes et al., 1998).

Disseminated coccidioidomycosis may be present in 1%–5% of individuals infected with *Coccidioides* spp., despite the absence of clinical or radiological signs of pulmonary involvement (Cox and Magee, 2004; Crum et al., 2004). The disseminated form most commonly affects immunosuppressed adult individuals, such as patients with AIDS or lymphomas, or transplanted patients (Galgiani et al., 2005). Fungal dissemination occurs by hematogenous and/or lymphatic pathways, and can reach several organs, such as skin, central nervous system, lymph nodes, bones, joints, and urogenital system. Disseminated coccidioidomycosis has a high mortality rate (Galgiani et al., 2005), so it requires accurate diagnosis and treatment. In Brazil, the Ministry of Health recommends differential diagnosis with visceral leishmaniasis (kalazar), especially in areas of occurrence of both diseases.

Primary cutaneous coccidioidomycosis occurs after traumatic inoculation of fungal structures and is associated with laboratory accidents (Cox and Magee, 2004). Dermatological manifestations include papules, nodules, and verrucous plaques, which may progress to the formation of ulcers and abscesses. Due to the clinical diversity of the lesions, cutaneous coccidioidomycosis can be confused with several other diseases, making laboratory confirmation necessary (Crum et al., 2004).

14.2 Collection and transport of clinical samples for mycological diagnosis

Since the specimens must be representative of the site of infection, several samples may be referred to the mycology laboratory depending on the clinical form of the disease. Thus, research of *Coccidioides* spp. can be performed on respiratory secretions, pleural fluid, cerebrospinal fluid (CSF), aspirated material from bone-joint lesions, and skin from biopsies. However, the most commonly investigated clinical specimens for laboratory diagnosis of coccidioidomycosis are those from the respiratory tree, such as sputum, and samples obtained by bronchoscopy: bronchoalveolar lavage and endobronchial and transbronchial biopsies. In addition, these samples usually show higher rates of fungal isolation compared to other sites, such as urine, blood, bone marrow, and CSF (Saubolle, 2007). Thus, samples from the respiratory tract will be highlighted in this chapter.

Clinical samples should be sent as soon as possible to the laboratory to prevent bacterial growth of the microbiota, especially in samples obtained from non-sterile sites. In addition, immediate transport aims to avoid the conversion of parasitic spherules to infective filamentous forms, which could pose serious health risks to laboratory workers. There is no need for special means of transport, but it is necessary to use sterile containers resistant to breakage and with hermetic seals. Biopsy material cannot be sent in formalin, and the use of isotonic saline is recommended.

The manipulation of clinical specimens from patients suspected of coccidioidomycosis requires laboratory infrastructure compatible with BSL-2. The manipulation of filamentous cultures of *Coccidioides* spp., however, requires BSL-3 practices, due to the high virulence of these organisms. It is important to emphasize that laboratories need to be alert to the adoption of strategies that avoid the conversion of the parasitic form to the infecting form in the biological samples being processed (such as the temporary stock at 4°C), as well as the possible occurrence of filaments in clinical samples. Some studies have revealed that type 2 diabetic patients with pulmonary coccidioidomycosis were four times more likely than non-diabetics to develop parasitic mycelial forms in respiratory samples (Muñoz-Hernández et al., 2008). Mycelial forms have been found in more than 50% of clinical samples from patients with cavitary chronic pulmonary coccidioidomycosis (Muñoz-Hernández et al., 2014), as well as in bronchoalveolar lavage samples (Helig & Giampoli, 2007), samples from patients with meningeal coccidioidomycosis (Hagman et al., 2000; Zepeda et al., 1998), and in ventriculoperitoneal shunts (Wages et al., 1995), as well as cutaneous lesions in disseminated disease (Kappel et al., 2007).

Coccidioides spp. are considered select agents of bioterrorism by the US government, and manipulation of cultures requires legal authorization and background checks of all employees in that country.

14.3 Laboratory processing

Intense mucoid sputum samples need to be digested with mucolytic agents (dithiothreitol or N-acetyl-1-cysteine) and then concentrated by centrifugation (1000–3000 g for at least 10 min). In general, it is recommended to concentrate all respiratory samples by centrifugation in order to maximize fungal isolation. For safety reasons, it is necessary to use break resistant flasks with hermetic seals; sample handling must occur at least 30 minutes after centrifugation of the tubes. It is suggested to analyze at least three sputum samples on consecutive days to exclude the diagnostic hypothesis of fungal infection.

14.3.1 Direct examination

For wet mounts, the clinical material is mixed with clarifying substance, such as 10% KOH or 10% NaOH, and allowed to stand for a few minutes in a humid chamber. Calcofluor white (CFW) fluorescent stain is indicated as the best dye for visualization of fungal structures in clinical specimens (Saubolle, 2007), although it requires proper equipment for fluorescence observation and requires well-trained personnel. CFW stains fatty materials nonspecifically (Saubolle, 2007). However, according to some authors, Papanicolaou stain is more sensitive than 10% KOH and CFW (Sarosi et al., 2001). The use of Gram staining for the diagnosis of coccidioidomycosis is controversial (Saubolle, 2007; Sutton, 2007) and should not be used as the sole screening method in suspect samples.

In direct microscopic examination, spherules of different sizes in various stages of maturation, up to 100 μm in diameter, and endospores with 2–5 μm in diameter are observed (Figure 14.1A and B). Ruptured spherules, spherules devoid of endospores, or free endospores and fragments of spherules may be seen (Figure 14.1C). Immature spherules (10–20 μm) lacking endospores or just free endospores can be confused with atypical forms of *Blastomyces dermatitidis* and yeast forms of *Histoplasma capsulatum*, *Candida glabrata*, or *Cryptococcus* species (Saubolle, 2007; Sutton, 2007). Atypical filamentous structures with arthroconidia are sometimes found in samples from patients with cavitary lung lesions (Brennan-Krohn et al., 2018; Kappel et al., 2007; Muñoz-Hernández et al., 2014). For sputum samples, the visualization of at least 25 polymorphonuclear leukocytes and fewer than ten epithelial cells per field, at 10× magnification, suggests material quality and indicates minimal contamination with saliva and oropharyngeal secretion.

14.3.2 Culture

Compared with other species of filamentous fungi, *Coccidioides* spp. are poor nutrient competitors, so it is recommended to use culture media containing selective agents—such as cycloheximide and chloramphenicol, gentamicin, streptomycin, or penicillin—for their isolation from samples that may have anemophilous filamentous fungi, yeasts, and/or bacteria. On the other hand, *Coccidioides* spp. do not show specific nutritional requirements, which allows

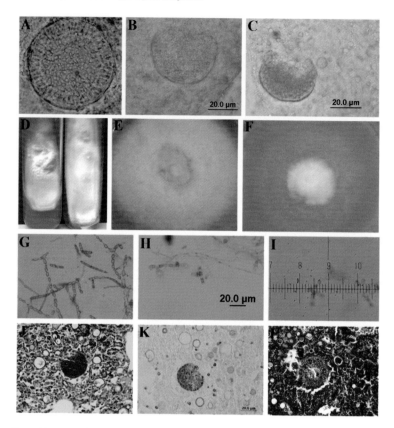

Figure 14.1   Morphological characteristics of *Coccidioides* spp. KOH mounts of sputa showing mature spherules (A, B) and ruptured spherules releasing endospores (C). Primary culture on potato agar dextrose (D), Sabouraud agar (E), and BHI agar (F). Colonies are apiculated, velvety (D, E), or cottony (F), white colored. Micromorphology is characterized by septate hyaline hyphae and arthroconidia spaced apart by barrel-shaped disjunctor cells (G–I). Wet mounts prepared with saline (G) or lactophenol cotton blue (H, I). Lung biopsy sections stained by periodic acid Schiff (J), Grocott methenamine silver (K), and Fontana Masson (L), showing mature parasitic structures.

their isolation—even unintentional—in different culture media. In addition to conventional mycological media—that is, Sabouraud agar, potato agar—the isolation of *Coccidioides* spp. can be performed on brain-heart infusion agar, sheep blood agar, or chocolate agar (Saubolle, 2007; Sutton, 2007), as well as on buffered charcoal-yeast extract agar and Bordet-Gengou agar, formulated for isolation of *Legionella* and *Bordetella*, respectively (Saubolle, 2007). Although limited by biosafety issues, culture is still considered the gold standard for the laboratory diagnosis of coccidioidomycosis.

Clinical specimens should be inoculated into hermetically sealed test tubes; biopsy material should be fragmented with a scalpel and never macerated. For safety reasons, suspected samples of coccidioidomycosis are not allowed to be cultured in Petri dishes. The test tubes should be incubated at 25°C–30°C and maintained for a minimum period of 7 days for visualization of the characteristic micromorphological structures. Although they are classified as dimorphic fungi, interconversion between parasitic and filamentous structures *in vitro* requires culturing strategies in special synthetic formulas and is beyond the scope of this work, but can be found elsewhere (Miyaji & Nishimura, 1985; Petkus et al., 1985; Sun & Huppert, 1976).

Primary cultures from clinical specimens show rapid development in approximately 5–10 days of incubation. Initially, the colonies are apiculated, glabrous, and white-colored. Microscopic analysis shows only septate hyaline hyphae. Usually after the fifth day of cultivation, morphological changes occur and the colonies are apiculated, velvety or cottony, white, cream, or gray-colored (Figure 14.1D and F) (Cordeiro et al., 2006). Tease-mounts for micromorphology examinations could be prepared in lactophenol cotton blue, and the coverslips must be sealed before manipulation (Sutton, 2007). Microscopic analysis of the mycelium reveals septate hyaline hyphae with approximately 3.5–5.0 µm width and interleaved arthroconidia spaced apart by barrel-shaped disjunctor cells (Figure 14.1G and I). Higher arthroconidia numbers are found after the sixth day of growth, when colonies are not membranous anymore. In addition, the dimensions of arthroconidia increase with time, varying from 5–7.5 µm and from 2.5–5 µm for length and width, respectively (Cordeiro et al., 2006).

14.4 Histopathological findings

Histopathological examination for *Coccidioides* spp. detection is an important diagnostic tool, especially when cultures cannot be performed. The histopathological diagnosis offers results faster than *in vitro* cultures and costs less than molecular and immunological techniques. It is recognized as providing presumptive or definitive diagnosis of the disease.

In immunocompetent patients, at the beginning of the infection, due to the large number of spherules and endospores, there is formation of granulomas with a high number of polymorphonuclear neutrophils, and to a lesser extent granulomatous cells. As the endospores mature, resulting in new spherules, the granulomatous response becomes predominant, with epithelioid cells, T and B lymphocytes, and multinucleated giant cells in the granulomas. Mixed inflammatory response occurs after release of endospores by mature beads, inducing a new influx of neutrophils. Eosinophils may be abundant, leading to the formation of the Splendore-Höeppli phenomenon around fungal structures.

The spherules and endospores can be evidenced by different dyes used in the histopathological analysis, such as periodic acid Schiff, Grocott methenamine silver, Fontana Masson (Figure 14.1J through L), and hematoxylin-eosin. Among these, Grocott methenamine silver is considered the most sensitive histological dye. Giemsa, Papanicolau, and mucicarmine are indicated as the least sensitive dyes in the histopathological diagnosis of coccidioidomycosis (Saubolle, 2007).

14.5 Molecular diagnosis

The laboratory diagnosis of coccidioidomycosis using molecular techniques is quite promising, since it allows the elimination of the risk in the handling filamentous cultures of *Coccidioides* spp., besides enabling detection of fungal DNA before seroconversion or in contaminated clinical samples and negative cultures (Bialek et al., 2004).

Protocols for molecular identification are mainly based on PCR (conventional or real-time systems) from clinical specimens (Binnicker et al., 2007; de Aguiar Cordeiro et al., 2007; Dizon et al., 2019; Saubolle et al., 2018) or cultures (Bialek et al., 2004; Tintelnot et al., 2007; Umeyama et al., 2006). Bronchoalveolar lavage fluid, bronchial washings, pleural fluid and sputum, and tissue samples (fresh, formalin-fixed, or paraffin-embedded) have been tested for molecular identification and details. Several molecular targets have been used for the diagnosis of coccidioidomycosis, with emphasis on the internal transcribed spacer (ITS) region of ribosomal DNA (Binnicker et al., 2007; Tintelnot et al., 2007), a coding sequence for the proline-rich antigen Ag2/PRA (Bialek et al., 2004; de Aguiar Cordeiro et al., 2007) and sequence of a genus-specific transposon (Saubolle et al., 2018) (Table 14.1). These tests have sensitivity and specificity above 95% on average.

However, it is important to note that the molecular diagnosis of coccidioidomycosis is expensive due to the high cost of equipment and supplies and is inaccessible to most microbiology laboratories in the cities of Latin America, where the disease is endemic. In addition, there are no commercial kits, so each laboratory should establish protocols, internal validation, and control tests for all stages of the diagnosis (DNA extraction and purification, target sequence amplification, and detection).

Table 14.1 PCR Assays for Identification of *Coccidioides* in Clinical Mycological Laboratories

Target	Primer Sequence (5'→3')	Reference
PRA gene	Cocci I GTACTATTAGGGAGGATAATCGTT	Bialek et al. (2004)
	Cocci II GGTGTCAACTGGGATGTCAAT	
	Cocci III ATCCCACCTTGCGCTGTATGTTCGA	
	Cocci IV GGAGACGGCTGGATTTTTTAACATG	
ITS rDNA	ITS C1CATCATAGCAAAAATCAAAC	Johnson et al. (2004)
	ITS C2 AGGCCCGTCCACACAAG	
Not informed	Coi9-1F TACGGTGTAATCCCGATACA	Umeyama et al. (2006)
	Coi9-1R GGTCTGAATGATCTGACGCA	
ITS2 rDNA	Forward CGAGGTCAAACCGGATA	Binnicker et al. (2007)[a]
	Reverse CCTTCAAGCACGGCTT	
	Anchor probe (3' fluorescein labeled)	
	GAGCGATGAAGTGATTTCCC	
	Donor probe (5' LC RED-640 labeled)	
	TACACTCAGACACCAGGAACT CG	
ITS rDNA	ITSv3 CAGTCTGAGCATCATAGC	Tintelnot et al. (2007)
	ITS4 TCCTCCGCTTATTGAIAIGC	

Abbreviations: PRA, Antigen rich in proline; ITS, internal transcribed spacer.
[a] Designed for RT-PCR.

14.6 Serological tests

The most relevant diagnostic methods for the detection of humoral response in coccidioidomycosis are gel immunodiffusion (ID) and enzyme immunoassay (EIA). Latex complement fixation and latex agglutination tests are also performed less frequently (Ampel, 2010; Saubolle, 2007). These tests are based on the detection of antibodies directed to tube precipitin (TP) and complement fixation (CF) antigens for the detection of immunoglobulin M (IgM) and immunoglobulin G (IgG), respectively (Martins et al., 1995; Pappagianis, 2001). Antigens TP and CF are obtained from the filtrate cultures of *C. immitis* in the mycelial phase, and, although subjected to purification, these antigenic preparations present cross-reactions with other systemic mycoses, such as histoplasmosis and blastomycosis. Despite this inconvenience, the use of the tests is valid for diagnosis and should be evaluated in a clinical-epidemiological approach.

Coccidioidal antibodies have no protective activity, but serve as tools for the diagnosis, prognosis, and therapeutic follow-up of patients (Pappagianis, 2001). Antibody screening can be performed in several clinical specimens, such as serum, CSF, pleural fluid, peritoneal fluid, and synovium (Pappagianis, 2001), depending on the standardization of each method. Tests can also be used to diagnose immunosuppressed patients, but in these cases there is a need to combine more than one diagnostic method (Blair et al., 2006; Johnson & Einstein, 2006).

After the onset of symptoms, the early detection of IgM can be performed between the first and third weeks, with positivity in 50% and 90% of the cases, respectively (Murthy & Blair, 2009). Approximately 90% of individuals show reactivity to IgG-directed tests around the fourth week after the onset of symptoms (Pappagianis, 2001; Saubolle, 2007). For IgM detection, the most commonly used methods in routine laboratories are gel immunodiffusion (IDTP) and enzyme immunoassay. IgG detection is performed by gel immunodiffusion (IDCF), complement fixation reaction, and EIA (Ampel, 2010). The detection of IgM or IgG can also be performed by the qualitative immunochromatographic method (lateral flow).

Gel immunodiffusion is a rapid qualitative or quantitative test that can provide presumptive diagnosis of coccidioidomycosis. Due to its facility, it is considered an excellent screening tool

(Blair et al., 2006). IgM antibody-directed ID becomes positive within 2 weeks of onset of symptoms in approximately 80% of cases, with rare occurrence up to 6 months after infection. In this case, the tests are qualitative and are not useful for antibody screening in CSF (Malo et al., 2014). In the test directed to IgG antibodies, the positive reaction is described as evidence of current or recent infection by *Coccidioides*, although some individuals show detectable levels of antibodies for more than 1 year after clinical cure (Pappagianis, 2001). As with other immunologic tests, negative results do not exclude the possibility of *Coccidioides* infection. In addition, cross-reactions can be seen in patients with other systemic mycoses, especially histoplasmosis (Pappagianis, 2001). Quantification of IgG titers by immunodiffusion may be performed in a number of clinical specimens, including CSF, and can be employed to monitor patient response to treatment (Malo et al., 2014). The test is usually employed to confirm EIA IgG results.

The complement fixation reaction is a qualitative or semi-quantitative assay of great value in the presumptive diagnosis of coccidioidomycosis and is generally positive in 75% of immunocompetent patients and 67% of immunocompromised patients (Pappagianis, 2001). By this technique, specific antibodies can be detected within 4–6 weeks after infection. The increase of the titres in sequential samples, as well as the presence of titres superior to 1:16, is suggestive of disseminated disease (Martins et al., 1995). Conversely, the reduction of these values reflects the resolution of the infectious condition. Since the complement fixation reaction is directed only to IgG antibody testing, the test should not be used alone for the diagnosis of coccidioidomycosis in the early acute stages, although it is considered the standard serological method for the diagnosis of meningeal forms of the disease (Pappagianis, 2001; Saubolle, 2007).

EIA is considered the most sensitive immunodiagnostic method, although the results directed to IgM antibodies are less specific and require confirmation by immunodiffusion or complement fixation (Saubolle, 2007). Usually, commercial tests are available in a panel for IgM and IgG detection.

In addition, the immunological diagnosis can be directed to the search for circulating fungal antigens, since in some patients the detection of these molecules precedes the detection of antibodies. This strategy can be of great importance in the presumptive diagnosis of the disease, especially in immunocompromised patients (Pappagianis, 2001). Available commercial tests allow the detection of galactomannan or Beta-1,3-glucan in urine, EDTA-treated serum, CSF, bronchoalveolar lavage, and other biological fluids. Cross-reactivity with other endemic fungi, such as *Histoplasma*, *Blastomyces*, or *Paracoccidioides*, has been described with galactomannan test antigen (Kuberski et al., 2007), as well as with *Aspergillus* and *Pneumocystis*, in the beta-1,3-glucan antigen test (Malo et al., 2014).

For the immunodiagnosis of coccidiodomyomycosis, commercial kits are available for the detection of antibodies (Immy Immuno Mycologics, USA; Miravista Diagnostics, USA; Meridian Bioscience, USA) and antigens (Miravista Diagnostics, USA; Associates of Cape Cod, Inc., USA) by ID, complement fixation, lateral flow, or EIA.

Despite the high costs of importing commercial kits and supplies, the immunological diagnosis of coccidioidomycosis has the advantage that it can be carried out in laboratories with biosafety level 2. However, the results obtained must always be evaluated in a broad clinical context for validation. In general, positive results are expected in paired serological tests (initial and follow-up) in patients with clinical signs and symptoms of the disease. Paired tests are considered positive when: (1) seroconversion occurs in samples tested at intervals of at least 3 weeks, (2) there is an increase in antibody titers in the samples, or (3) additional positivity occurs in more than one serological test (Malo et al., 2014).

14.7 Conclusion

Coccidioidomycosis is a neglected disease in South America, and many cases are expected to be misdiagnosis with tuberculosis—an endemic disease in many Latin America countries. In Brazil, an epidemiological link between coccidioidomycosis and hunting armadillos had been extensively described. Handling of *Coccidioides* cultures requires BSL-3 facilities and well trained personnel. Clinical laboratories in endemic areas should provide at least immunological tests for presumptive diagnosis of coccidioidomycosis.

References

Ampel, N. M. 2010. The diagnosis of coccidioidomycosis. *F1000 Med Rep* 2: 2.

Bialek, R., Kern, J., Herrmann, T. et al. 2004. PCR assays for identification of *Coccidioides posadasii* based on the nucleotide sequence of the antigen 2/proline-rich antigen. *J Clin Microbiol* 42: 778–783.

Binnicker, M. J., Buckwalter, S. P., Eisberner, J. J. et al. 2007. Detection of *Coccidioides* species in clinical specimens by real-time PCR. *J Clin Microbiol* 45: 173–178.

Blair, J. E., Coakley, B., Santelli, A. C., Hentz, J. G., Wengenack, N. L. 2006. Serologic testing for symptomatic coccidioidomycosis in immunocompetent and immunosuppressed hosts. *Mycopathologia* 162: 317–324.

Brennan-Krohn, T. Yoon, E., Nishino, M., Kirby, J. E., Riedel, S. 2018. Arthroconidia in lung tissue: An unusual histopathological finding in pulmonary coccidioidomycosis. *Hum Pathol* 71: 55–59.

Castañeda-Godoy, R., Laniado-Laborín, R. 2002. Coexistencia de tuberculosis y coccidioidomicosis. Presentación de dos casos clínicos. *Rev Inst Nal Enf Resp Mex* 15: 98–101.

Chiller, T. M., Galgiani, J. N., Stevens, D. A. 2003. Coccidioidomycosis. *Infect Dis Clin North Am* 17: 41–57.

Cordeiro, R. A., Brilhante, R. S., Rocha, M. F. et al. 2006. Phenotypic characterization and ecological features of *Coccidioides* spp. from Northeast Brazil. *Med Mycol* 44: 631–639.

Cox, R. A., Magee, D. M. 2004. Coccidioidomycosis: Host response and vaccine development. *Clin Microbiol Rev* 17: 804–839.

Crum, N. F., Lederman, E. R., Stafford, C. M., Parrish, J. S., Wallace, M. R. 2004. Coccidioidomycosis: A descriptive survey of a reemerging disease. Clinical characteristics and current controversies. *Medicine* 83: 149–175.

de Aguiar Cordeiro, R., Nogueira Brilhante, R. S., Gadelha Rocha, M. F., Araújo Moura, F. E., Pires de Camargo, Z., Costa Sidrim, J. J. 2007. Rapid diagnosis of coccidioidomycosis by nested PCR assay of sputum. *Clin Microbiol Infect* 13: 449–451.

de Hoog, G. S., Guarro, J., Gené, J., Figueras, M. J. 2000. *Atlas of Clinical Fungi*. 2nd Edition. Utrecht, The Netherlands: Centraalbureau voor Schimmelcultures/Universitat Rovira i Virgili.

Dizon, D., Mitchell, M., Dizon, B., Libke, R., Peterson, M. W. 2019. The utility of real-time polymerase chain reaction in detecting *Coccidioides immitis* among clinical specimens in the Central California San Joaquin Valley. *Med Mycol* 57: 688–693.

Galgiani, J. N., Ampel, N. M., Blair, J. E. et al. 2005. Coccidioidomycosis. *Clin Infect Dis* 41: 1217–1223.

Gomes, O. M., Serrano, R. R., Pradel, H. O. et al. 1978. Coccidioidomicose pulmonar. Primeiro caso nacional. *Rev Assoc Med Bras* 24: 167–168.

Hagman, H. M., Madnick, E. G., D'Agostino, A. N. et al. 2000. Hyphal forms in the central nervous system of patients with coccidioidomycosis. *Clin Infect Dis* 30: 349–353.

Hector, R. F., Laniado-Laborin, R. 2005. Coccidioidomycosis: A fungal disease of the Americas. *PLoS Med* 2: e2.

Helig, D., Giampoli, E. J. 2007. Mycelial form of *Coccidioides immitis* diagnosed in bronchoalveolar lavage. *Diagn Cytopathol* 35: 535–536.

Johnson, R. H., Einstein, H. E. 2006. Coccidioidal meningitis. *Clin Infect Dis* 42: 103–107.

Johnson, S. M., Simmons, K. A., Pappagianis, D. 2004. Amplification of coccidioidal DNA in clinical specimens by PCR. *J Clin Microbiol* 42:1982–1985.

Kappel, S. T., Wu, J. J., Hillman, J. D., Linden, K. G. 2007. Histopathologic findings of disseminated coccidioidomycosis with hyphae. *Arch Dermatol* 143: 548–549.

Kirkland, T. N., Fierer, J. 2018. *Coccidioides immitis* and *posadasii*; A review of their biology, genomics, pathogenesis, and host immunity. *Virulence* 9: 1426–1435.

Kuberski, T., Myers, R., Wheat, L. J. et al. 2007. Diagnosis of coccidioidomycosis by antigen detection using cross-reaction with a *Histoplasma* antigen. *Clin Infect Dis* 44: 50–54.

Malo, J., Luraschi-Monjagatta, C., Wolk, D. M., Thompson, R., Hage, C. A., Knox, K. S. 2014. Update on the diagnosis of pulmonary coccidioidomycosis. *Ann Am Thorac Soc* 11: 243–253.

Martins, T. B., Jaskowski, T. D., Mouritsen, C. L., Hill, H. R. 1995. Comparison of commercially available enzyme immunoassay with traditional serological tests for detection of antibodies to *Coccidioides immitis*. *J Clin Microbiol* 33: 940–943.

Miyaji, M., Nishimura, K. 1985. Conversion of *Coccidioides immitis* from a mycelial form to spherules using the "agar-implantation method". *Mycopathologia* 90: 121–123.

Moraes, M. A. P., Martins, R. L. M., Leal, I. I. R., Rocha, I. S., Medeiros Junior, P. 1998. Coccidioidomicose: Novo caso brasileiro. *Rev Soc Bras Med Trop* 31: 559–562.

Muñoz-Hernández, B., Martínez-Rivera, M. A., Palma-Cortés, G., Tapia-Díaz, A., Manjarrez Zavala, M. E. 2008. Mycelial forms of *Coccidioides* spp. in the parasitic phase associated to pulmonary coccidioidomycosis with type 2 diabetes mellitus. *Eur J Clin Microbiol Infect Dis* 27: 813–820.

Muñoz-Hernández, B., Palma-Cortés, G., Cabello-Gutiérrez, C., Martínez-Rivera, M. A. 2014. Parasitic polymorphism of *Coccidioides* spp. *BMC Infect Dis* 14: 213.

Murthy, M. H., Blair, J. E. 2009. Coccidioidomycosis. *Curr Fungal Infect Rep* 3: 7–14.

Pappagianis, D. 2001. Serologic studies in coccidioidomycosis. *Semin Respir Infect* 16: 242–250.

Petkus, A. F., Baum, L. L., Ellis, R. B., Stern, M., Danley, D. L. 1985. Coccidioides immitis in continuous culture. *J Clin Microbiol* 22: 165–167.

Sarosi, G. A., Lawrence, J. P., Smith, D. K., Thomas, A., Hobohm, D. W., Kelley, P. C. 2001. Rapid diagnostic evaluation of bronchial washings in patients with suspected coccidioidomycosis. *Semin Respir Infect* 16: 238–241.

Saubolle, M. A. 2007. Laboratory aspects in the diagnosis of coccidioidomycosis. *Ann N Y Acad Sci* 1111: 301–314.

Saubolle, M. A., Wojack, B. R., Wertheimer, A. M., Fuayagem, A. Z., Young, S., Koeneman, B. A. 2018. Multicenter clinical validation of a cartridge-based real-time PCR system for detection of *Coccidioides* spp. in lower respiratory specimens. *J Clin Microbiol* 24: e01277–17.

Schneider, E., Hajjeh, R. A., Spiegel, R. A. et al. 1997. A coccidioidomycosis outbreak following the Northridge, Calif, earthquake. *JAMA* 277: 904–908.

Sun, S. H., Huppert, M. 1976. A cytological study of morphogenesis in *Coccidioides immitis*. *Sabouraudia* 14: 185–198.

Sutton, D. A. 2007. Diagnosis of coccidioidomycosis by culture: Safety considerations, traditional methods, and susceptibility testing. *Ann N Y Acad Sci* 1111: 315–325.

Tintelnot, K., De Hoog, G. S., Antweiler, E. et al. 2007. Taxonomic and diagnostic markers for identification of *Coccidioides immitis* and *Coccidioides posadasii*. *Med Mycol* 45: 385–393.

Umeyama, T., Sano, A., Kamei, K., Niimi, M., Nishimura, K., Uehara, Y. 2006. Novel approach to designing primers for identification and distinction of the human pathogenic fungi *Coccidioides immitis* and *Coccidioides posadasii* by PCR amplification. *J Clin Microbiol* 44: 1859–1862.

Wages, D. S., Helfend, L., Finkle, H. 1995. *Coccidioides immitis* presenting as a hyphal form in a ventriculoperitoneal shunt. *Arch Pathol Lab Med* 119: 91–93.

Zepeda, M. R., Kobayashi, G. K., Appleman, M. D., Navarro, A. 1998. *Coccidioides immitis* presenting as a hyphal form in cerebrospinal fluid. *J Natl Med Assoc* 90: 435–436.

15
Pneumocystis jirovecii

Rosely Maria Zancopé-Oliveira, Fernando Almeida-Silva,
Rodrigo de Almeida Paes, and Mauro de Medeiros Muniz

Contents

15.1 Introduction

Pneumocystis pneumonia (PJP) in humans is caused by the fungus *Pneumocystis jirovecii*, which affects mainly immunocompromised patients. PJP, formerly known as pneumonia by *Pneumocystis carinii* (PCP), was an early indicator of the human immunodeficiency virus (HIV) epidemic, and in the 1980s occurred in 70%–80% of AIDS patients (CDC 1982; Masur et al., 1981). However, the incidence of PJP in AIDS has decreased as a result of early HIV diagnosis, better antiretroviral therapy, and the use of prophylaxis measures. PJP manifestations are mainly associated with the respiratory tract and symptoms are generally nonspecific. Diagnosis is hampered by the inability to culture the organism and based on microscopic examination of respiratory samples, x-ray patterns, or clinical presentation. New assays can assist in the diagnosis and even aid with the emergence of resistant infections (White et al., 2017).

15.2 Collection, transport, and processing of samples

Respiratory samples are necessary for the diagnosis of PJP. These include bronchoalveolar lavage (BAL), induced sputum, and lung biopsies, among others. Samples must be collected by a physician following the specific medical guidelines and under aseptic conditions. They should be sent to the laboratory as soon as possible, for a faster diagnosis that can improve patient's survival, especially in those not infected by the human immunodeficiency virus (Asai et al., 2012).

15.3 Microscopic examination and culture

Pneumocystis jirovecii cannot be observed in the direct examination of samples using potassium hydroxide (KOH 10%) mounts and also cannot be cultured in the mycologic routine media (Sabouraud and Mycosel agar). However, these techniques must be performed for a differential diagnosis with

Figure 15.1   Honeycomb structures compatible with *Pneumocystis jirovecii* in bronchoalveolar lavage. (A) Silver methenamine stain, magnification of 1000×. (B) Direct immunofluorescence, magnification of 400×. (C) Immunohistochemistry, magnification of 400×.

other respiratory infections and to detect mixed infections with other pathogenic fungi, such as *Cryptococcus neoformans* and *Histoplasma capsulatum* (Almeida-Silva et al., 2016).

The gold standard for the diagnosis of PJP remains the histological and microscopic identification of ascus and trophic forms using different stains (White et al., 2017). The microscopic examination of *P. jirovecii* structures in respiratory samples can be achieved by staining with the methenamine silver stain (Figure 15.1A) or with the Giemsa stain. The former stains only the cell wall of cystic forms, whereas the later stains the nuclei of all *Pneumocystis* life stages of the fungus. Toluidine blue stains nucleic acids and calcofluor white stains chitin and cellulose. Although frequently applied to the direct examination of *Pneumocystis*, they are not specific for this fungus. The sensitivity of these methods has a broad range of variation. In comparison of the three staining methods, sensitivities range from 73.8%, 76.9%, and 48.4%, respectively for calcofluor white, Grocott-Gomori (methenamine silver stain), and Diff-Quik (modified Wrights-Giemsa), whereas specificity was higher than 99%. The sensitivity and specificity can be affected by the fungal burden into the lungs and by the observer accuracy and expertise (White et al., 2017).

Although *Pneumocystis* lacks growth into mycological routine media, a three-dimensional air-liquid interface culture system using CuFi-8 (a differentiated pseudostratified airway epithelial cell line) that supports *P. jirovecii* growth obtained from bronchoalveolar lavage has been described in the last few years. The complexity of the system makes its use in the laboratorial routine impractical; however, it will make possible several research studies on the fungal cell cycle, drug sensitivity, growth, and virulence (Schildgen et al., 2014).

15.4 Immunological test for *P. jirovecii* antibody detection

Considering that the standard laboratory diagnosis of PJP is based on microscopic visualization of stained *P. jirovecii*, several factors can hinder the conventional laboratory results, including patient clinical conditions; the need for invasive techniques and their associated risk of complications; accuracy of the microscopic approach, depending on observer skills; and chemoprophylaxis, leading to a low burden of *P. jirovecii*, especially in HIV-infected patients taking highly active antiretroviral therapy (Song et al., 2016). Therefore, the development of alternative serological tests would allow the diagnosis of PJP using just blood samples, undoubtedly a less invasive sample. Nevertheless, serological tools are also used for microscopic detection of *P. jirovecii* in respiratory tract specimens. Both methodologies, that is, immunofluorescence (Figure 15.1B) and immunohistochemistry assays (Figure 15.1C), use monoclonal antibodies directed against specific antigens of *P. jirovecii* (Song et al., 2016).

The performance of conventional staining has been supplanted by immunofluorescence (IF) microscopy using anti-*P. jirovecii* monoclonal antibodies (Alanio et al., 2016). Several commercial kits presenting different sensitivities are available for routine diagnostic (White et al., 2017). In those kits, the detection's reagent contains fluorescein isothiocyanate (FITC)-labeled monoclonal antibodies directed against cell wall and matrix antigens of *P. jirovecii*, which attach to the fungi present in the clinical specimen examined under fluorescence microscope. As a result, fungal cells will present a bright apple-green fluorescence for the characteristic morphology of *P. jirovecii*, and the background in the specimen is counterstained from orange to red. The sensitivities for immunofluorescent antibodies against surface antigens of *P. jirovecii* are higher than those using dye-staining methods (Kaur et al., 2015).

The type and quality of the specimen used could affect the assay performance. For instance, in a study comparing both IF and conventional staining on sputum and BAL, the sensitivity was lower in all tests when sputum was analyzed (Cregan et al., 1990).

In order to improve the PJP diagnosis, promising studies using recombinant antigens of *P. jirovecii* and antibody detection methods, such as immunoenzymatic assays, have shown potential application in the diagnosis and epidemiological studies of PJP during the last years (Blount et al., 2012; Djawe et al., 2010, 2013; Gingo et al., 2011). For several years, the main studied *Pneumocystis* antigen was the major surface glycoprotein (Msg), a highly glycosylated protein of 90–120 kDa encoded by multiple genes with an estimation of 50–100 copies per cell. According to Stringer (2005), this molecule contains shared and species-specific epitopes capable of eliciting humoral and cellular protective immune responses and plays a central role in the interaction of *Pneumocystis* with its host.

Recently, Tomás et al. (2016) produced a multi-epitope recombinant synthetic antigen (RSA) of *P. jirovecii* Msg and applied it to an indirect enzyme-linked immunosorbent assay (ELISA) method for the detection of anti-*P. jirovecii* IgM and IgG antibodies. The test presented 100% and 80.8% sensitivity and specificity, respectively, when associated with clinical diagnosis criteria. This new method may be used as a screening test for PJP, decreasing the need for biological specimens obtained by invasive techniques, which is a major benefit to the patient's care and an improvement in the clinical management of the disease.

15.5 Serological biomarkers: (1–3)-β-D-glucan, lactate dehydrogenase, Krebs von den Lungen-6, and S-adenosylmethionine

Many serological methods have been studied using some biomarkers of the disease. Actually, such metabolite levels are not specific to *P. jirovecii* infection. However, high levels could be found in clinical samples from PJP such as lactate dehydrogenase (LDH); the structural component of fungi—(1–3)-β-D-glucan (BDG); and Krebs von den Lungen-6 antigen (KL-6), a mucinous high-molecular weight glycoprotein expressed on type 2 pneumonocytes. However, they can also be found in other pathologies involving cell damage (Esteves et al., 2015).

Low serological levels of S-adenosylmethionine (AdoMet/SAM), have been related to PJP. AdoMet/SAM is a naturally occurring compound found in almost every tissue and fluid in the body, and it is involved in many important processes and plays a role in the immune response (Esteves et al., 2015). The measurement of AdoMet/SAM levels in serum was proposed as a useful biochemical alternative and promising diagnostic test for PJP in patients infected with HIV, demonstrating a sensitivity and specificity of >90% (Skelly et al., 2008). However, the utility of AdoMet/SAM as an indicator for PJP was evaluated in immunocompromised HIV-negative patients from the Netherlands, in which its measurement in serum failed to discriminate between patients with and without PJP, revealing very low diagnostic robustness (de Boer et al., 2011).

BDG is the most used biological marker in the diagnosis of PJP. A positive result alone, however, cannot be considered diagnostic of PJP, due to the wide-range cross-reaction with other fungal pathogens. Non-infective factors could also affect the specificity of the test (White et al., 2017). However, this assay is now widely accepted when results are interpreted along with radiological findings.

Several studies addressing the association of some biomarkers have been reported. The combination of BDG testing in association with LDH levels has been successfully evaluated for the diagnosis of PJP (Esteves et al., 2014). Posteriorly, a robust work has shown information about serum levels of BDG, LDH, KL-6, and AdoMet/SAM. The combination of BDG and KL-6 presented the best performance, with 94% sensitivity and 90% specificity. The sensitivity (97%) of the test was increased in the presence of BDG and LDH; however, there was a decrease to 72% in the specificity (Esteves et al., 2015). Although these tests are less efficient than the reference standard classic methods based on microscopic or molecular detection of *P. jirovecii*, the blood BDG/KL-6 test may put forward a less onerous procedure for PJP diagnosis, providing a major benefit for the patient's care. Clinical context must be correlated with the results of the BDG/KL-6 combination test, which could be used as a preliminary screening test in patients with primary PJP suspicion or as an alternative diagnostic procedure in patients suffering from respiratory insufficiency or in children, avoiding the associated risk of complications due to the use of bronchoscopy.

A serious problem involving PJP diagnosis concerns the distinction between *P. jirovecii* colonization and true PJP. Unfortunately, serum biomarker levels are not able to distinguish those clinical conditions, that is, none of them has established a definitive optimal cut-off limit to be used safely up to now (Damiani et al., 2013; Matsumura et al., 2012; Shimizu et al., 2009; Tasaka et al., 2014).

15.6 Molecular diagnosis of *Pneumocystis jirovecii* pneumonia

Molecular diagnosis of PJP is based on the detection of *P. jirovecii* DNA in clinical samples. However, direct PCR detection of *P. jirovecii* DNA does not necessarily constitute a diagnosis of infection, because the microorganism may be present in its latent form. Nevertheless, the excellent performance for PJP PCR in upper and lower respiratory tract specimens, when testing HIV-infected and non-HIV-infected patients, has been demonstrated by several studies focusing on the comparison of different techniques (Fan et al., 2013; Lu et al., 2011; Summah et al., 2013). The evidence suggests that the main factor involved in sensitivity and specificity of molecular assays is the gene target chosen. In general, ribosomal DNA (ITS, 18SrRNA, 5SrRNA regions) is frequently the target of different PCR formats.

However, *P. jirovecii* has only one copy of the gene encoding the ribosomal RNA, explaining why techniques that use ITS region as diagnostic target present low efficiency (Giuntoli et al., 1994). The selection of multicopy genes such as *mtLSUrRNA* and *Msg* may increase the sensitivity, as shown in a study comparing these two targets (Fischer et al., 2001). mtLSUrRNA is fully involved in basic metabolic processes; this region has a high degree of genetic conservation (Beard et al., 2000). Besides, there are several mitochondria in *P. jirovecii* cells increasing sensitivity of the nested PCR. For this reason, partial sequences of Msg and mtLSUrRNA are the most suitable molecular target for the detection of *P. jirovecii*.

Clinical samples are also important for the sensitivity and specificity of the test. Among the respiratory specimens, better results of PCR-based methods are found in BAL specimens, with 98.3% and 91.0% sensitivity and specificity, respectively (Fan et al., 2013). It has been suggested that, even with moderate sensitivity and specificity, molecular tests should be performed on less invasive samples, since BAL is difficult to obtain (Chawla et al., 2011).

However, the application of PCR-based methods on less invasive samples has lower efficiency because the number of organisms in these samples is insufficient to yield a detectable DNA concentration after fungal lysis (Matos et al., 2011).

15.6.1 Conventional polymerase chain reaction

When performed on sputum or BAL, the reported sensitivity and specificity of a conventional PCR is around 100% (Azoulay et al., 2009). The assays frequently target the ribosomal DNA (ITS, 18SrRNA, 5SrRNA regions), thymidylate synthase (*TS*), and *P. jirovecii Msg* genes, as well as *mtLSUrRNA* of *P. jirovecii*. Detection of *P. jirovecii* through targeting *mtLSUrRNA* using primers pAZ102-H and pAZ102-E has 98.3% and 91.0% sensitivity and specificity, respectively. In 2014, molecular detection of *P. jirovecii* by conventional PCR was improved by the development of a new forward primer, named pH207, used with the initial reverse primer (pAZ102-E). This advance revealed a new amplification assay with cost efficiency, turnaround time, and reduced contamination risks (Chabé et al., 2014). Conventional PCR was initially a useful technique; however, its results when compared with nested PCR are less satisfactory (Chabé et al., 2014).

15.6.2 Nested polymerase chain reaction

This assay is characterized by the use of an initial first amplification as the template for the second amplification round. This type of reaction increases the sensitivity without decreasing specificity. A study comparing nine molecular methods targeting different regions was performed, and after the review of clinical histology/cytology records to resolve discrepancies, the nested PCR to *mtLSUrRNA* was the most sensitive method (Roberts et al., 2006). Compared with single rounds of PCR, nested PCR is more indicated for diagnosis, because conventional PCR is sometimes not sufficient to amplify low amounts of DNA in clinical samples. However, this technique is time consuming compared to conventional PCR (Fan et al., 2013).

15.6.3 Real-time polymerase chain reaction (qualitative polymerase chain reaction)

Currently, the use of conventional PCR amplification systems has been replaced by real-time PCR (quantitative PCR—qPCR) platforms (White et al., 2017). Several reports using qPCR show the high significance of this quantitative method and the unique molecular method able to differentiate between colonization and infection, a crucial advance for the survival of PJP patients (Song et al., 2016). For this goal, the fungal burden based on the PCR cycle threshold should be evaluated (Alanio et al., 2011; Botterel et al., 2012; Fauchier et al., 2016; Lu et al., 2011). The interpretation of the fungal burden significance should be associated with the underlying conditions of the patient and quality of the sample assayed. In addition, sensitivity and specificity higher than 95% should be considered when setting thresholds to confirm or exclude disease (White et al., 2017). As an example, the Fast Track Diagnostics (FTD) *Pneumocystis* PCR kit targeting the *mtLSUrRNA* gene of *P. jirovecii* was evaluated for this purpose. Respiratory specimens were examined using both microscopy and PCR assay. The median copy numbers in BAL were significantly different in the PJP and colonization groups (1.35×10^8/mL vs. 1.45×10^5/mL, $P < 0.0001$). Lower and upper cutoff values of 3.9×10^5 copies/mL and 3.2×10^6 copies/mL demonstrated the differentiation between PJP and colonization, showing that the FTD *Pneumocystis* PCR kit had good performance and represents an alternative method to diagnose *P. jirovecii* infections (Hoarau et al., 2017).

Due to the large number of PCRs developed for the diagnosis of PJP, it has been recommended that commercial tests, well standardized and with instructions about their interpretation, should be used in routine laboratories. For instance, in a study developed by Sasso et al. (2016), comparing the *P. jirovecii* FRT PCR kit (AmpliSens), MycAssay *Pneumocystis* (Myconostica), and real-time PCR *Pneumocystis jirovecii* (Bio-Evolution) in proven/probable PJP, sensitivities of 100%, 100%, and 95%, and specificities of 83%, 93%, and 100%, respectively, were observed. Excellent sample concordance between *P. jirovecii* FRT PCR kit (AmpliSens) and MycAssay *Pneumocystis* (Myconostica) was also determined.

15.7 Combination testing

Currently, the question of whether combining tests would increase the sensitivity and specificity of the diagnosis of PJP remains without answer. A previous study comparing four methods (Giemsa, IF, single-round PCR, and nested PCR) showed that nested PCR targeting *mtLSUrRNA* significantly increased the positivity, improving the clinical practice (El-Seidi et al., 2008). Posteriorly, current guidelines suggest a diagnostic algorithm involving qPCR and IF testing of BAL in patients with clinical suspicion of PJP in adult hematology populations (Alanio et al., 2016). If both are positive, a diagnosis of PJP is confirmed. If the PCR is positive, but IF is negative, diagnosis is made if high fungal burdens are detected. For low burdens, additional BDG testing is recommended. If the PCR is negative, but IF is positive, then the result is considered technically inconsistent (Alanio et al., 2016).Other combinations include PJP PCR and BDG testing on BAL when the pulmonary burden is low with BDG testing (Damiani et al., 2013). When BAL samples are not available, BDG testing of serum is recommended where negativity can be used to exclude PJP, but positivity should be confirmed by PCR (or IF) testing of less invasive respiratory samples. However, given the current limitations of PCR (possible detection of colonization) and BDG (detection of other fungal species) testing, there may be a preference to combine PCR with microscopy to confirm or exclude PJP when testing deeper respiratory specimens (Alanio et al., 2016).

15.8 Summary and conclusions

The main global clinical approach is driven to reduce the practice of empirical therapy, especially in low-middle income countries, in patients with severe respiratory failure and mainly in children, in whom the execution of invasive techniques such as bronchoscopy is not easy to accomplish. In view of a myriad of methodologies for the diagnosis of PCP, the best results will always be those that merge clinical, radiological, and laboratory findings.

References

Alanio, A., Desoubeaux, G., Sarfati, C. et al. 2011. Real-time PCR assay-based strategy for differentiation between active *Pneumocystis jirovecii* pneumonia and colonization in immunocompromised patients. *Clin Microbiol Infect* 17: 1531–1537.

Alanio, A., Hauser, P. M., Lagrou, K. et al. 2016. ECIL guidelines for the diagnosis of *Pneumocystis jirovecii* pneumonia in patients with hematological malignancies and stem cell transplant recipients. *J Antimicrob Chemother* 71: 2386–2396.

Almeida-Silva, F., Damasceno, L. S., Serna, M. J. et al. 2016. Multiple opportunistic fungal infections in an individual with severe HIV disease: A case report. *Rev Iberoam Micol* 33: 118–121.

Asai, N., Motojima, S., Ohkuni, Y. et al. 2012. Early diagnosis and treatment are crucial for the survival of *Pneumocystis* pneumonia patients without human immunodeficiency virus infection. *J Infect Chemother* 18: 898–905.

Azoulay, É., Bergeron, A., Chevret, S., Bele, N., Schlemmer, B., Menotti, J. 2009. Polymerase chain reaction for diagnosing *Pneumocystis* pneumonia in non-HIV immunocompromised patients with pulmonary infiltrates. *Chest* 135: 655–661.

Beard, C. B., Carter, J. L., Keely, S. P. et al. 2000. Genetic variation in *Pneumocystis carinii* isolates from different geographic regions: Implications for transmission. *Emerg Infec Dis* 6: 265–272.

Blount, R. J., Jarlsberg, L. G., Daly, K. R. et al. 2012. Serologic responses to recombinant *Pneumocystis jirovecii* major surface glycoprotein among Uganda patients with respiratory symptoms. *PLOS ONE* 7: e51545.

Botterel, F., Cabaret, O., Foulet, F., Cordonnier, C., Costa, J. M., Bretagne, S. 2012. Clinical significance of quantifying *Pneumocystis jirovecii* DNA by using real-time PCR in bronchoalveolar lavage fluid from immunocompromised patients. *J Clin Microbiol* 50: 227–231.

Centers for Disease Control and Prevention (CDC). 1982. A cluster of Kaposi's sarcoma and *Pneumocystis carinii* pneumonia among homosexual male residents of Los Angeles and Orange Counties, California. *MMWR Morb Mortal Wkly Rep* 31: 305–307.

Chabé, M., Khalife, S., Gantois, N., Even, G., Audebert, C. 2014. An improved single-round PCR leads to rapid and highly sensitive detection of *Pneumocystis* spp. *Med Mycol* 52: 841–846.

Chawla, K., Martena, S., Gurung, B., Mukhopadhyay, C., Varghese, G. K., Bairy, I. 2011. Role of PCR for diagnosing *Pneumocystis jirovecii* pneumonia in HIV-infected individuals in a tertiary care hospital in India. *Indian J Pathol Microbiol* 54: 326–329.

Cregan, P., Yamamoto, A., Lum, A., VanDerHeide, T., MacDonald, M., Pulliam, L. 1990. Comparison of four methods for rapid detection of *Pneumocystis carinii* in respiratory specimens. *J Clin Microbiol* 28: 2432–2436.

Damiani, C., Le Gal, S., Da Costa, C., Virmaux, M., Nevez, G., Totet, A. 2013. Combined quantification of pulmonary *Pneumocystis jirovecii* DNA and serum (1/3)-β-D-glucan for differential diagnosis of *Pneumocystis* pneumonia and *Pneumocystis* colonization. *J Clin Microbiol* 51: 3380–3388.

de Boer, M. G., Gelinck, L. B., van Zelst, B. D. et al. 2011. β-D-glucan and S-adenosylmethionine serum levels for the diagnosis of *Pneumocystis* pneumonia in HIV-negative patients: A prospective study. *J Infect* 62: 93–100.

Djawe, K., Daly, K. R., Levin, L., Zar, H. J., Walzer, P. D. 2013. Humoral immune responses to *Pneumocystis jirovecii* antigens in HIV infected and uninfected young children with *Pneumocystis* pneumonia. *PLOS ONE* 8: e82783.

Djawe, K., Huang, L., Daly, K. R. et al. 2010. Serum antibody levels to the *Pneumocystis jirovecii* major surface glycoprotein in the diagnosis of *P. jirovecii* pneumonia in HIV+ patients. *PLOS ONE* 5: e14259.

El-Seidi, E. A., El-Hodaky, S. K., Ouda, N. H. et al. 2008. Comparison of different methods to diagnose *Pneumocystis jirovecii* pneumonia in children with haematological malignancies. *Egyp J Med Microbiol* 17: 151–159.

Esteves, F., Calé, S. S., Badura, R. et al. 2015. Diagnosis of *Pneumocystis* pneumonia: Evaluation of four serologic biomarkers. *Clin Microbiol Infect* 21: 379 e1−e10.

Esteves, F., Lee, C. H., de Sousa, B. et al. 2014. (1-3)-beta-D-glucan in association with lactate dehydrogenase as biomarkers of *Pneumocystis* pneumonia (PcP) in HIV-infected patients. *Eur J Clin Microbiol Infect Dis* 33(7): 1173–1180.

Fan, L. C., Lu, H. W., Cheng, K. B., Li, H. P., Xu, J. F. 2013. Evaluation of PCR in bronchoalveolar lavage fluid for diagnosis of *Pneumocystis jirovecii* pneumonia: A bivariate meta-analysis and systematic review. *PLOS ONE* 8: e73099.

Fauchier, T., Hasseine, L., Gari-Toussaint, M., Casanova, V., Marty, P. M., Pomares, C. 2016. Detection of *Pneumocystis jirovecii* by quantitative PCR to differentiate colonization and pneumonia in immunocompromised HIV-positive and HIV-negative patients. *J Clin Microbiol* 54: 1487–1495.

Fischer, S., Gill, V. J., Kovacs, J. et al. 2001. The use of oral washes to diagnose *Pneumocystis carinii* pneumonia: A blinded prospective study using a polymerase chain reaction–based detection system. *J Infect Dis* 184: 1485–1488.

Gingo, M. R., Lucht, L., Daly, K. R. et al. 2011. Serologic responses to *Pneumocystis* proteins in human immunodeficiency virus patients with and without *Pneumocystis jirovecii* pneumonia. *J Acquir Immune Defic Syndr* 57: 190–196.

Giuntoli, D., Stringer, S. L., Stringer, J. R. 1994. Extraordinarily low number of ribosomal RNA genes in *P. carinii*. *J Eukaryot Microbiol* 41: 88S.

Hoarau, G., Le Gal, S., Zunic, P. et al. 2017. Evaluation of quantitative FTD-*Pneumocystis jirovecii* kit for *pneumocystis* infection diagnosis. *Diagn Microbiol Infect Dis* 89: 212–217.

Kaur, R., Wadhwa, A., Bhalla, P., Dhakad, M. S. 2015. *Pneumocystis* pneumonia in HIV patients: A diagnostic challenge till date. *Med Mycol* 53: 587–592.

Lu, Y., Ling, G., Qiang, C. et al. 2011. PCR diagnosis of *Pneumocystis* pneumonia: A bivariate meta-analysis. *J Clin Microbiol* 49: 4361–4363.

Masur, H., Michelis, M. A., Greene, J. B. et al. 1981. An outbreak of community-acquired *Pneumocystis* carinii pneumonia: Initial manifestation of cellular immune dysfunction. *N Engl J Med* 10 305: 1431–1438.

Matos, O., Costa, M. C., Lundgren, B., Caldeira, L., Aguiar, P., Antunes, F. 2011. Effect of oral washes on the diagnosis of *Pneumocystis* carinii pneumonia with a low parasite burden and on detection of organisms in subclinical infections. *Eur J Clin Microbiol Infect Dis.* 20: 573–575.

Matsumura, Y., Ito, Y., Iinuma, Y. et al. 2012. Quantitative real-time PCR and the (1/3)-β-D-glucan assay for differentiation between *Pneumocystis jirovecii* pneumonia and colonization. *Clin Microbiol Infect* 18: 591–597.

Roberts, F. J. L., Liebowitz, L. D., Chalkley, L. J. 2006. Polymerase chain reaction detection of *Pneumocystis jirovecii*: Evaluation of 9 assays. *Diagn Microbiol Infect Dis* 58: 385–392.

Sasso, M., Chastang-Dumas, E., Bastide, S. et al. 2016. Performances of four real-time PCR assays for diagnosis of *Pneumocystis jirovecii* pneumonia. *J Clin Microbiol* 54: 625–630.

Schildgen, V., Mai, S., Khalfaoui, S. et al. 2014. *Pneumocystis jirovecii* can be productively cultured in differentiated CuFi-8 airway cells. *MBio* 5: e01186–14.

Shimizu, Y., Sunaga, N., Dobashi, K. et al. 2009. Serum markers in interstitial pneumonia with and without *Pneumocystis jirovecii* colonization: A prospective study. *BMC Infect Dis* 9: 47.

Skelly, M. J., Holzman, R. S., Merali, S. 2008. S-Adenosylmethionine levels in the diagnosis of *Pneumocystis* carinii pneumonia in patients with HIV infection. *Clin Infect Dis* 46: 467–471.

Song, Y., Ren, Y., Wang, X., Li, R. 2016. Recent advances in the diagnosis of *Pneumocystis* pneumonia. *Med Mycol J* 57: E111–E116.

Stringer, J. R. 2005. Surface antigens. In Walzer, P. D., Cushion, M. T. editors *Pneumocystis* pneumonia (3rd Edition) Marcel Dekker, Inc., New York, USA, [chapter 4].

Summah, H., Zhu, Y. G., Falagas, M. E., Vouloumanou, E. K., Qu, J. M. 2013. Use of real-time polymerase chain reaction for the diagnosis of *Pneumocystis* pneumonia in immunocompromised patients: A meta-analysis. *Chin Med J (Engl).* 126: 1965–1973.

Tasaka, S., Kobayashi, S., Yagi, K. et al. 2014. Serum (1/3) β-d-glucan assay for discrimination between *Pneumocystis jirovecii* pneumonia and colonization. *J Infect Chemother* 20: 678–681.

Tomás, A. L., Cardoso, F., Esteves, F., Matos, O. 2016. Serological diagnosis of pneumocystosis: Production of a synthetic recombinant antigen for immunodetection of *Pneumocystis jirovecii*. *Sci Rep* 6: 36287.

White, P. L., Backx, M., Barnes, R. A. 2017. Diagnosis and management of *Pneumocystis jirovecii* infection. *Exp Rev Anti-Infect Ther* 15: 435–447.

Index

Pocket Guides to Biomedical Sciences

Series Editor
Lijuan Yuan

A Guide to AIDS, *by Omar Bagasra and Donald Gene Pace*

Tumors and Cancers: Central and Peripheral Nervous Systems, *by Dongyou Liu*

A Guide to Bioethics, *by Emmanuel A. Kornyo*

Tumors and Cancers: Head – Neck – Heart – Lung – Gut, *by Dongyou Liu*

Tumors and Cancers: Skin – Soft Tissue – Bone – Urogenitals, *by Dongyou Liu*

Tumors and Cancers: Endocrine Glands – Blood – Marrow – Lymph, *by Dongyou Liu*

A Guide to Cancer: Origins and Revelations, *by Melford John*

Pocket Guide to Bacterial Infections, *edited by K. Balamurugan*

A Beginner's Guide to Using Open Access Data, *by Saif Aldeen Saleh Airyalat and Shaher Momani*

Pocket Guide to Mycological Diagnosis, *edited by Rossana de Aguiar Cordeiro*

For more information about this series, please visit: https://www.crcpress.com/
Pocket-Guides-to-Biomedical-Sciences/book-series/CRCPOCGUITOB

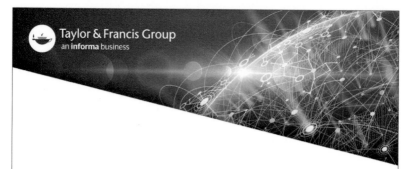